Festschrift Column

Regular Features

ASC Column

Book Reviews

The artist for this issue is Pille Bunnell.
Issue poet: Kathleen Forsythe.

Cover Art

Bunnell, P. (2015). *Living Lines*. Un-retouched photograph.

CYBERNETICS & HUMAN KNOWING
A Journal of Second-Order Cybernetics, Autopoiesis & Cyber-Semiotics
ISSN: 0907-0877

Cybernetics and Human Knowing is a quarterly international multi- and trans-disciplinary journal focusing on second-order cybernetics and cybersemiotic approaches.

The journal is devoted to the new understandings of the self-organizing processes of information in human knowing that have arisen through the cybernetics of cybernetics, or second order cybernetics its relation and relevance to other interdisciplinary approaches such as C.S. Peirce's semiotics. This new development within the area of knowledge-directed processes is a non-disciplinary approach. Through the concept of self-reference it explores: cognition, communication and languaging in all of its manifestations; our understanding of organization and information in human, artificial and natural systems; and our understanding of understanding within the natural and social sciences, humanities, information and library science, and in social practices like design, education, organization, teaching, therapy, art, management and politics. Because of the interdisciplinary character articles are written in such a way that people from other domains can understand them. Articles from practitioners will be accepted in a special section. All articles are peer-reviewed.

Subscription Information

Price: Individual £69.30. Institutional: £152.25 (online); £186.90 (online & print). 50% discount on full set of back volumes. Payment by cheque in £UK (pay Imprint Academic) to PO Box 200, Exeter EX5 5HY, UK; Visa/Mastercard/Amex.
email: sandra@imprint.co.uks

Editor in Chief: Søren Brier, Professor in semiotics at the Department of International Culture and Communication Studies attached to the Centre for Language, Cognition, and Mentality, Copenhagen Business School, Dalgas Have 15, DK-2000 Frederiksberg, Denmark, Tel: +45 38153246. sb.ikk@cbs.dk

Editor: Jeanette Bopry, Instructional Sciences, National Institute of Education, 1 Nanyang Walk, Singapore 637616. jeanette.bopry@gmail.com

Associate editor: Dr. Paul Cobley, Reader in Communications, London Metropolitan University, 31 Jewry Street, London EC3N 2EY. p.cobley@londonmet.ac.uk

Managing editor: Phillip Guddemi, The Union Institute and University, Sacramento CA, USA. pguddemi@well.com

Joint art and website editor: Claudia Jacques
cj@claudiajacques.org

C&HK is indexed/abstracted in *Cabell's Journal* and *Psycinfo*
Journal homepage: www.chkjournal.com
Full text: www.ingenta.com/journals/browse/imp

Copyright: It is a condition of acceptance by the editor of a typescript for publication that the publisher automatically acquires the English language copyright of the typescript throughout the world, and that translations explicitly mention *Cybernetics & Human Knowing* as original source.

Book Reviews: Publishers are invited to submit books for review to the Editor.

Instructions to Authors: To facilitate editorial work and to enhance the uniformity of presentation, authors are requested to send a file of the paper to the Editor on e-mail. If the paper is accepted after refereeing then to prepare the contribution in accordance with the stylesheet information at www.chkjournal.org

Manuscripts will not be returned except for editorial reasons. The language of publication is English. The following information should be provided on the first page: the title, the author's name and full address, a title not exceeding 40 characters including spaces and a summary/ abstract in English not exceeding 200 words. Please use italics for emphasis, quotations, etc. Email to: sbr.lpf@cbs.dk

Drawings. Drawings, graphs, figures and tables must be reproducible originals. They should be presented on separate sheets. Authors will be charged if illustrations have to be re-drawn.

Style. CHK has selected the style of the APA (*Publication Manual of the American Psychological Association*, 5th edition) because this style is commonly used by social scientists, cognitive scientists, and educators. The APA website contains information about the correct citation of electronic sources. The APA Publication Manual is available from booksellers. The Editors reserve the right to correct, or to have corrected, non-native English prose, but the authors should not expect this service. The journal has adopted U.S.English usage as its norm (this does not apply to other native users of English). For full APA style informations see: apastyle.apa.org

Accepted WP systems:
MS Word and rtf.

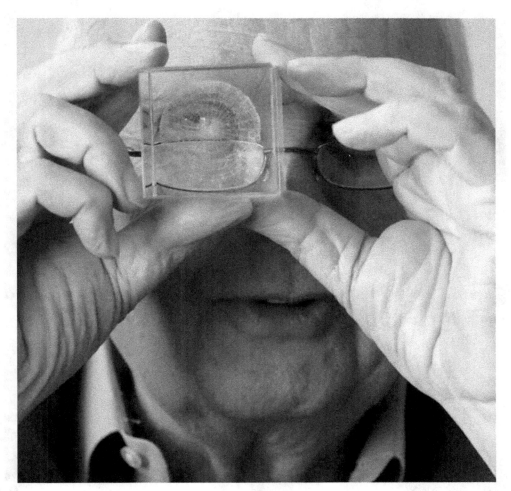

Ranulph Glanville (Photo: Delmar Mavignier, vimeo.com/channels/zerospaces)

Bunnell, P. (2010). *Nude Trunk*. Un-retouched photograph.

Cybernetics and Human Knowing. Vol. 22 (2015), nos. 2-3, pp. 7-10

Foreword: Ranulph Glanville and How to Live the Cybernetics of Unknowing
A Festschrift Celebration of the Influence of a Researcher

Phillip Guddemi, Søren Brier and Louis H. Kauffman

It is not customary to use journals for festschrifts. But we have made an exception with our columnist—and contributor to the journal through the years—Ranulph Glanville. Since our journal began, his work has created great academic interest among our readers, and we wanted researchers from the many areas he touched with his work to reflect on the nature and impact of this unusual engaged thinker, who combined breadth and depth in a very original way throughout his life and practice.

This tribute issue to the effects of Ranulph Glanville's life work began as a celebration of his life. Søren Brier and Lou Kauffman sent out the call for contributions on October 28 of last year, while Ranulph was still alive, though they were aware of his terminal illness. Allow us to reprint most of the call for contributions' first paragraph:

> Ranulph Glanville is about to retire from his work for the Cybernetics & Human Knowing journal ... He will also retire from great parts of his enormous fan of work for associations such as the ASC, conferences and other journals. We therefore thought we would commemorate his gigantic work with a festschrift commenting on the significance of what he has done.

This issue is the result of that call, but of course made much more poignant by Ranulph's subsequent untimely death.

The issue begins with three short introductory pieces. Fittingly the first is from Ranulph's wife, Aartje Hulstein, who has been his discussion partner in the production of many of the columns. The difficult art Ranulph practiced in his columns was that of being deep and yet easily understandable, and close to real life interaction, in a rather short text. She notes Ranulph's pleasure in learning of this festschrift, and describes the spirit in which he did his work, with an emphasis on his crafting of his regular columns for this journal.

Following this piece is an overview article by Søren Brier, which is both an introduction to Ranulph's work and a description of the author's engagement in the inspired discussions that went on from the first draft of the columns to the final version. Concluding this introductory section is a short remembrance by Mary Catherine Bateson of Ranulph and his work for the American Society for Cybernetics for which he was president his last years.

Ranulph's Ph.D. dissertation on Objects was intended by him as the foundation of his career, but it is little known even within cybernetic circles. Albert Müller has undertaken to provide us with an explication of this thesis, how it came to be and how Ranulph used its ideas in his later work.

Following Albert Müller's article are three pieces of great intellectual depth in which Ranulph's work is assessed and contextualized *in the round* (a theatrical term suggesting from many perspectives). These are philosophical pieces which are as challenging as the work deserves. Karl Müller aptly names his article, *De Profundis*, for reasons he explains in his text, and in it he illustrates Ranulph's place in the constellation of second-order cybernetics, as well as explaining his theory of Objects and his application of cybernetics in the fields of communication, learning and design. This has been a concept which has been difficult to fit into the dominant discussion. Karl Müller's piece is an interesting sociological and philosophical reflection on the interaction between a radically new creative thinking researcher and the fields he was touching—and how difficult it is, sociologically and existentially, to formulate new insights and interdisciplinary directions in established fields of research and practices.

Dirk Baecker also touches on these issues learnedly but here in the context of the incipient cybernetics of the work of Martin Heidegger, who inspired by Schelling saw cybernetics as the fulfilling of modern science's full aspiration.

Bernard Scott explains the more direct connection of Ranulph's work with that of his teacher Gordon Pask, who developed conversation theory. It was a theory which Ranulph not only promoted but also related to his theory of Objects (and one he would have liked to have seen much more discussion about in this journal).

The pieces which follow give some emphasis to the idea of design, which was the focus of Ranulph's teaching for many decades. Hugh Dubberly and Paul Pangaro, in a concise yet profound way, demonstrate the depth and rigor of a concept of design deeply inspired by Ranulph's theory and practice. They begin, fittingly, with their conversation with Ranulph at one of his last presentations, that of the RSD3 2014 Symposium in Oslo, and they show how Ranulph's thought was still developing and refocusing even in the face of his illness.

In the following piece, Robert Martin relates the idea of design to second-order cybernetics with a focus on composition and music—indeed many people may not know that Ranulph was an experimental composer and musician. Robert Martin's piece is followed by several in which the concept of design relates to its more usual association with architecture. Gerard de Zeeuw and Rolf Hughes relate how research in architecture, both observational and non-observational, fit with Ranulph's cybernetic approach. Ben Sweeting shows the intimate correspondence between Ranulph's theory of design and its inspiration from conversation theory, in which a cybernetic practice informs both and provides both with an ethics. This is also the subject of Christiane Herr's article, which also deals with radical constructivism as an approach Ranulph found valuable in his work with design.

The issue concludes with a number of pieces which focus on Ranulph's teaching presence in his last years and particularly his work as President of the American

Society for Cybernetics. Notwithstanding the somewhat personal reflections that comprise these pieces, and their relationship with the frustrating cybernetics of governance, they all retain the rigor which relates these matters to theory, specifically second-order cybernetic theory and the cybernetics of design.

There are two pieces with nearly the same title, "What I Learned from Ranulph Glanville." One of them is from a former president of the ASC and the other is from the new incumbent president who has followed Ranulph in the position. The former President is Larry Richards and he describes Ranulph's clarity of thought, commitment to listening, quiet determination, conversation (theory and practice), and concept of design. Larry Richards concludes with a conversation he would like to have had with Ranulph about the theory of government. The new President is Michael Lissack, who gives tribute to Ranulph by expanding upon a conversation the two of them had after the transfer of power (or position) from the one to the other. The conversation was on the subject of stridency and polarization, a topic on which Michael Lissack expands theoretically at some length.

The next two pieces also focus on specific conversations with Ranulph. Thomas Fischer in "Designing Together" describes specific conversations on the topic of mutual design, and an example is given of the logo and diagram for the 2013 ASC conference. The conversation here includes much that took place in email, including after Ranulph became ill. Philip Baron in a piece called "Glanville's Consistency" departs from the usual academic format to show Ranulph in a direct relation of conversation, including a dialogue about therapy that became pivotal in Philip Baron's life. It shows Ranulph not only in theory but also as a person.

Finally there is a short reflection on "My Time with Ranulph Glanville" by the youngest contributor, Thomas Fischer's and Christiane Herr's daughter Lily—with some help from her parents. It is an existential view of the person Ranulph, seen through the eyes of a young child.

Lou Kauffman's regular column for this issue is his own version of a tribute to Ranulph, one which gives the Cookie and Parabel treatment—readers of the column will know these to be Kauffman's interlocutory alter egos—to a joint paper written by Ranulph Glanville and Francisco Varela, entitled "Your Inside is Out and Your Outside is In." The treatment is based on G. Spencer-Brown's *Laws of Form*.

This ends the festschrift.

We have also included an ASC column by Robert Martin. In it, as he states, he considers the failure of second-order cybernetics, radical constructivism, and the biology of condition to be fully accepted in science, and considers the opportunities that still exist for these ways of thinking in the cultural and intellectual world.

Also there are two book reviews from Phillip Guddemi. The first is about the new book by Ronald R. Kline, *The Cybernetics Moment, or Why We Call Our Age the Information Age*. Ronald Kline is the Bovay Professor in History and Ethics of Engineering at Cornell University. His book is, in part, an ambitious recounting of the history of cybernetics, beginning with Weiner and the Macy Conferences. But counterpointed to this is a study of how the idea of an information age has had such great appeal that it has in many ways overshadowed the cybernetics from which it was born.

The second book review discusses a book from the biosemiotician (and novelist) Victoria N. Alexander. The book's title, *The Biologist's Mistress*, refers to a comment attributed to the eminent biologist J. B. S. Haldane about teleology: "Biology cannot live without her but is unwilling to be seen with her in public." Dr. Alexander has chosen to identify openly as a teleologist and subsumes much creative thinking about complexity, systems, and biosemiotics under that category. She also discusses the history of teleology and how it can contribute to a theory of aesthetics and art.

For this issue, our Festschrift to Ranulph Glanville, the featured artist is Pille Bunnell, a systems ecologist and second-order cybernetician. Dr. Bunnell has been serving on the editorial board of the journal since she initiated an ASC column in 1999, the same year she began her three-year term as president for ASC. She served the journal as editor for the ASC column (1999 to 2012), and as art editor (2004 to 2012). Her connection with the ASC community led to many friendships and collaborations, not least of which was a deep and enduring friendship with Ranulph Glanville, whom she continued to encourage during his nearly decade long leadership, inclusive of two terms as president of the ASC.

We wish to express appreciation and respect for Glanville in part by recognizing his support for Pille's many contributions to our journal and the field of cybernetics. We also reveal a little-known side of her accomplishments by publishing herein a sampling of her luminous photographs of the natural world, which mirror her scholarship as they "explore the ramifications of reflections as they alter how we humans see ourselves and how we relate to each other and the world around us."

Bunnell's intimate relationship with the natural world as well as understanding of complex systems is revealed in her photographs. Her images show her delight in the designs found in the colors, values, shapes and textures created by water, earth, fungus, wood, and so forth. They also reflect the depth of her knowledge as a systems scientist and ecologist.

Poetry has been provided by Kathleen Forsythe. Photos of Ranulph were provided by Delmar Mavignier and Christiane M. Herr.

Bunnell, P. (2010). *Life Fell In*. Un-retouched photograph.

Cybernetics and Human Knowing. Vol. 22 (2015), nos. 2-3, pp. 11-12

Living Between Cybernetics Columns

Aartje Hulstein[1]

When Søren Brier and Lou Kauffman suggested a festschrift for Ranulph after he had decided he would stop writing the columns in *Cybernetics and Human Knowing* at the end of 2014, Ranulph was very pleased.

I am very pleased to see so many people commenting on Ranulph's work, and sad at the same time as Ranulph is not here anymore to enjoy this special issue and comment on it.

Ranulph often felt people did not appreciate his work and was surprised when one of the reactions to his diagnosis and illness was an increased interest in his work and what he had to offer. He wanted to write all that he had to say and increased his effort.

I am grateful for all the visitors we had in the past year, who made it possible for Ranulph to continue with his work, develop it further and talk about his work and write the last column.

Another project he was delighted about was the filming the Royal College of Art decided to do, to capture some of the "Ranulphness" for future generations.

A very special moment was the last lecture Ranulph gave in Oslo at the end of October 2014. Tim Jachna and Thomas Fischer came to Oslo to record this and to hand over the award Ranulph and I received from the ASC at the last conference. We could not attend due to treatment, but Ranulph joined by Skype.

Søren Brier asked me to write something for this festschrift, as cybernetics has played such a big role in our life.

When Ranulph and I met in 1994 in Amsterdam he had just started to write the columns for *Cybernetics and Human Knowing* and they have been part of our life together right up till the end. Cybernetics in the form of conferences, meeting people, the American Society for Cybernetics and papers to be written was very much part of our life and work together.

Ranulph not only talked about cybernetics, but also tried to live it. He would choose a concept and think about it. When the concept became clearer he would start to share with me what he was thinking about and we would explore it together. I often asked for clarification and how I could use it in my work, our life together.

I learned to observe better, to see what it did when I was aware of the observing. Learning that I saw the world differently from everyone else, and that made me more interested in how children and specially the disabled students I worked with composed their view of the world. If only I could see the world as they did for one moment, how would this change the way I treated them physically.

---

1. Email: aartje@glanville.co.uk

This for me opened up a whole new world of interest in my work as a pediatric physiotherapist. I would discuss my observations with Ranulph and people we visited. Richard Jung was one of the people who played a crucial role in this, explaining a different kind of intelligence, one that is in the doing and can only be grasped by reflection. Those conversations often resulted in a different approach and interaction with the teachers and students I worked with. We tried to make movement part of the educational process and also to add lots of fun to therapy.

After exploring the concept, for the columns or a paper, the writing would begin, a period of immense concentration. Writing the way Ranulph did was a creative act, he would try to simplify, make the words flow and let people experience what he was writing about.

The writings, but also his lectures, always became journeys, Ranulph took people and showed them how he saw the world and how that view could help others to understand the world differently. I often read the first drafts of the columns and papers and asked more questions. The next step would be that Søren Brier or other colleagues became involved, a conversation by email started and Ranulph would continue to rework the paper. In the end the result of those conversations was published.

Ranulph always worked in a conversational way with me and all the others he met and worked with. It did not matter whether he traveled the world, worked at universities with students and staff, gave keynote lectures, became president of the ASC and developed the new style conferences or met friends and relatives. An enormous curiosity how others thought and saw the world and engage with that was part of what made Ranulph so special. I had the privilege of observing it and taking part at the same time.

It is a credit to Ranulph and how he explained the thinking of second-order cybernetics to me that it has become so close to my heart as a way of understanding the world and how to live in it. It still helps me to deal with life and see the possibilities and joy it offers.

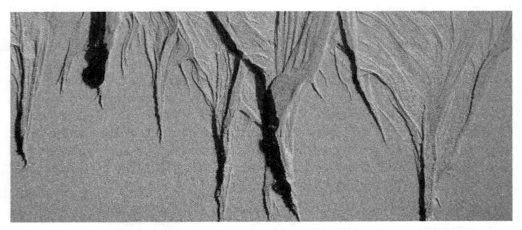

Bunnell, P. (2006). *Sandy Glyphs*. Un-retouched photograph.

Cybernetics and Human Knowing. Vol. 22 (2015), nos. 2-3, pp. 13-17

Ranulph Glanville: The Cybernetician of the Black Box of Second-order Unknowing

Søren Brier[1]

Seven years ago I wrote a little article called Ranulph Glanville: "The Cybernetician of Ignorance" (Brier, 2008). I had the famous book *Docta Ignoranta* by Nicolaus Cusanus about learned ignorance in mind, when I coined the title. I was thinking of a similarity between radical constructivism and the adage that inspired Cusanus' title, Pseudo-Dionysius the Areopagite's advice to his reader to "strive upwards unknowingly." In spite of all the knowledge we have gathered since then, it is still pretty much the situation we are in and it is a realization that is at the core of Ranulph Glanville's work.

Being pretty unknowing myself about many of Ranulph's published works, my earlier article was based on the evaluation work of Mary Catherine Bateson, Dirk Baecker and Stephen Gage, contributing to Brunel University's awarding the Doctor of Science (DSc) to Ranulph Glanville. As these evaluations were not published, I was graciously allowed by the authors to integrate their work into an article of my own to which I added a few comments on my own co-work with Ranulph in writing a column for this journal over the last twenty years. The last of these columns were published in the previous issue of *Cybernetics & Human Knowing* 22(1) (Brier & Kauffman, 2013).

These columns were collected and republished as one of the books in the great book project by edition echoraum which collected Glanville's published works in a three volume set entitled *The Black Box* as part of the series Complexity / Design / Society. Karl Müller explains more about this huge project in his article in this festschrift.

But the story I really wanted to tell is that when I—on request—a couple of years later put this article on Academia.edu for free reading, it very quickly turned out to be the most read of my articles ever on that webpage, not only on that year but for many years. Again and again people have been searching for it on the net because of Ranulph's great influence in so many areas and because this paper gives such a good overview of his work thanks to the three referees.

In his last years Ranulph—when we discussed his retirement from contributing two columns a year to CHK—doubted that his work had any broad impact. I then showed him the statistics of this paper for a bit of comfort and he was much surprised. I promised him on the basis of that to make a festschrift for him when he retired from

---

1. Søren Brier is Professor in the semiotics of information, cognition and communication science, department of International Business Communication, Copenhagen Business School, Dalgas Have 15, 2000 Frederiksberg, Copenhagen, Denmark. Email: sb.ikk@cbs.dk

his lifetime work in CHK, as I was convinced that the interest in it would be enormous.

The present text is unfortunately post mortem as Ranulph was in and out of hospital while we were working on his last column, and finally did not come back. But before that period I did manage to tell him during the last month of his life that the articles to this issue were pouring in from all over the globe where he had travelled and lectured so extensively in the last decades of his life; and this seemed to give some kind of comfort to him.

I have written several papers to 70 years birthdays festschrifts in recent years, and the feeling of being alone with one's ideas and efforts is quite common among famous interdisciplinary trail blazers. I think it is partly because breaking a new interdisciplinary path means that you do not really belong to any community. You do not have obvious colleagues and you pretty much have to create your own institutions—be they journals or societies. Ranulph Glanville contributed to the birth of, and sustained, many journals and societies and edited festschrifts and proceedings for many researchers and conferences. He was always helping and organizing, like Thomas Sebeok in semiotics. He was creating fields of inquiry and knitting networks all over the globe—in the last years not least as president of the ASC. He, much in collaboration with Louis H. Kauffman, certainly did a lot to uphold this journal that is now miraculously on its 22nd volume, floating between disciplines and societies upheld by independent interdisciplinary academics like him. CHK is a bumblebee kept in the air supported by a network of dedicated (unpaid) free-spirited academics and a wonderful publisher.

As an academic, Ranulph Glanville was amazingly idealistic and non-disciplinary in his dedication to working in the spaces between disciplines as well as between universities. He simply gave up his regular position at The Portsmouth Polytechnic School of Architectural design in 1996 when he found the institutional frames counter-productive. From then on he worked in a variety of part-time regular posts and ad hoc engagements all around the globe. This way of life seems to be the last place of existence for free academics in our post-industrial cognitive capitalist and public management society. The price of this freedom was excessive work and travel combined with global connectivity and immediate awareness. Contrary to many of my nationally-based colleagues with a steady job, I could always reach Ranulph on the net in a day or so.

We interacted in happy disagreement partly based in our different points of departure, his coming from cybernetic architecture and his apprenticeship with Gordon Pask and I from behavioural psycho-biology engaged in Gregory Bateson's work, which through my work on the Danish Journal *Paradigma* got connected to the ASC and Maturana's and von Foerster's work. The physicist Peder Voetmann Christiansen turned me on to the work of the great American pragmaticist C. S. Peirce, an engagement which through interchanges with Jesper Hoffmeyer, Claus Emmeche, Mogens Kielstrup, Kalevi Kull, and Fredrik Stjernfelt turned into biosemiotics. Anyway, over the years, Ranulph and I continued to engage in discussion whenever he produced a column; sometimes I would involve Lou Kauffman, to draw on his expertise of Spencer-Brown as I was having similar discussions with him whenever he made his contribution. Dirk Baecker, who also had ongoing exchanges with Ranulph, is also a researcher I had exchanges with over the years encouraging his publications with CHK as these have turned out to be of great value to the sort of knowledge processes the journal has wanted to promote. So the three of us were in intense enlightening discussion over the many years Ranulph contributed to the journal and since Louis Kauffman joined us as a second columnist.

We had an inter- and transdisciplinary interest in the foundations of knowledge in common, but we were coming from very different places, keeping us enough apart to require ongoing discussion. Ranulph Glanville with his architectural design teaching background, Louis Kauffman with his logic and mathematical insight, and finally Dirk Baecker in a social communicative philosophical space deeply influenced by Niklas Luhmann's work on which he is an authority. Both Baecker and I were in dialogue with another Luhmann inspired researcher that has done a lot for keeping the quality of the journal's dialogue up; namely my local colleague Ole Thyssen, with whom I worked in the Danish Academy of Applied Philosophy. Thyssen was responsible for CHK's contact with Luhmann in the last years of his life.

No doubt it was Pierce's semiotic philosophy that kept us somewhat apart, though both Louis Kauffman and Dirk Baecker had some access to his thinking. Baecker's background in the strong German philosophical tradition also shows in his article in this issue, as well as it does in Karl Müller's. In many ways it was Spencer-Brown's work that held the group together. In a couple of issues, with Louis Kauffmann, we explored the similarities and differences between Peirce's and Spencer-Brown's metaphysics of mathematics and logic. In the previous issue John Levi Martin's article on the relation between G. Spencer-Brown's early work and C.S. Peirce's approach to statistics continues this line of inquiry in a most interesting way.

My discussions with Ranulph Glanville focused very much on the lack of a fully reflected phenomenological viewpoint in second-order cybernetics and autopoiesis theory, starting with a critique of Bateson's "difference that makes a difference" for a cybernetic mind, and continuing into Maturana and Luhmann's work on autopoiesis. I claimed that the cybernetic mind of Bateson, as well as Maturana and Luhmann's autopoiesis and Von Foerster's Eigenforms, are not theoretically grounded in the experiential mind (Brier, 1992). What distinguishes Peirce's semiotics from cybernetics—even second-order cybernetics—is its phenomenological philosophical grounding. This point of view brought me into (healthy) disagreement with many of my Batesonian colleagues who have published in this journal.

In his very profound paper on Ranulph's work in this issue: *De Profundis: Ranulph Glanville's Transcendental Framework for Second-Order Cybernetics*, Karl Müller philosophizes on the radical constructivist movement and why Ranulph Glanville was not allowed into its hall of fame, as that movement swept over most of second-order cybernetics. I am honoured to be mentioned here, but like Maturana, von Foerster and Luhmann, I am not sure I am a radical enough to be called a radical constructivist since my main teaching effort the last 30 years has been in the philosophy of science. I still cling to a concept of truth (Brier & Kauffman, 2013) and realism through Peirce's pragmaticist semiotic fallibilism with its empirically founded never ending truth-finding process in a community of ideally engaged researchers looking for truth (Misak, 1995).[2]

But for Ranulph Glanville—like so many other cyberneticians —semiotics was an uninteresting field and they never enjoyed the profoundness of Peirce's triadic semiotic pragmaticism, though when I discussed Ranulph's last column with him I realized how close he came to Peirce's concept of semiotic objects, the semiotics net and the growth of symbols (Peirce, 1982-) in his own discussion of a new concept of "objects as wholes" and parts as "wholes with a role."

All in all Ranulph was also not very interested in discussing traditional philosophy of science and metaphysics since he considered his own groundwork on objects as a way out of the impasses of the development of traditional philosophy. In that way he was much like my semiotic and philosophy of science-oriented colleague John Deely, who in his *Four Ages of Understanding: The First Postmodern Survey of Philosophy from Ancient Times to the Turn of the Twenty-First Century* declared Charles Sanders Peirce to be the first true post-modern. Thus there are good reasons to dwell on the similarities between Spencer-Brown and Peirce's work.

In spite of Ranulph Glanville's reluctance to discuss traditional philosophy's way of formulating our problems of understanding, the knowledge and communication processes leading to truth in his last column comes very close to Leibniz's monadology, with its self-observing and autopoietic wholes that are always connected in an overall systemic harmony. When he writes about the glue that holds the wholes together, he reminded me again of Peirce's continuity philosophy, called synechism. With Peirce, Ranulph Glanville also shared a profound distinction between objects and things in the usual meaning of the word. This distinction is thoroughly explored by John Deely in his book *Purely Objective Reality*, because both Peirce and Deely prefer the pre-modern philosophical usage of *subject*, which makes it almost synonymous with thing in itself and like Ranulph Glanville tend to use *object* only in reference to the object of a sign.

This is why Deely has referred to Peirce as the first one who broke with the Cartesian tradition that dominated (and still dominates a lot of) modern philosophy, philosophy of science and much thinking in the sciences in the area of brain science and cognitive research. Thus it also makes Ranulph a true post-modern philosopher and composer. As Peirce says (W6:37) there are two subjects that are occult and mysterious for inter- and transdisciplinary science: One is the power of nature that brings about the result of experiments in the form of causality in the material dominated universe. This is mysterious because it is unobservable in itself, but never the less we are forced by this power to admit that there must be a system of regular

---

2.   See also my dialog with Louis Kauffman called "Nothing But the Truth," in the proceedings of Dirk Baecker's conference on an aspect of Spencer-Brown's work published in *CHK, 21*[1-2] (Brier & Kauffman, 2013).

relations between the causes and conditions constituting the experiment and the result of it, and we can only guess what this regular relation is in itself. It is what Peirce christened as the category of Thirdness.

The other "occult and mysterious" phenomenon is the power that connects the conditions of the mathematician's diagram with the relations he observes in it. You might say that the diagram is the relations that are observable in it. Note also the role of the diagram in games like Chess, where the position of the pieces on the board is the diagram and the player must contemplate that diagram for the relations implicit in it as a locus of processes that can emerge from it. Like physical causality, the necessity of mathematical (and logical) reasoning is no less compelling than physical causality, even though mathematical objects are imagined, and no more directly observable than physical causality is. Still "all reasoning involves observation" (W6:37). There are interesting similarities between Von Foerster's idea of Eigenforms and Pierce's idea of interactions between the forms of objects.

Like Peirce, Ranulph Glanville survived and produced partly outside the traditional institutions of the nation state academia. It took nearly 50 years after Peirce's death before serious work trying to understand the wholeness of his vision of human knowing and communication was made, and the work and discussions have continued up to the recent 100-year anniversary of his death. There are many intellectual advances connected to the position of a free roaming academic. You are freer of bureaucratic, economic and political influences. Not least Pierre Bourdieu (1988) has warned against the way the political power of the state sneaks in and forms the concepts with which we investigate our own culture. But this position can also be a hindrance of getting the results of you work out far enough around the globe. Fortunately this will not be the fate of Ranulph Glanville's work as his most important works recently have been gathered and published by edition echoraum, making the study of his legacy so much easier.

Acknowledgement

Thanks to Louis H. Kauffman for productive critical input.

References

Bourdieu, P. (1988). *Homo Academicus* (P. Collier, Trans.). Redwood City, CA: Stanford University Press.

Brier, S. (1992). Information and consciousness: A critique of the mechanistic concept of information. *Cybernetics & Human Knowing, 1*(2/3).

Brier, S. (2008). Ranulph Glanville: The cybernetician of ignorance. *Cybernetics & Human Knowing, 15*(1), 81-89.

Brier, S., & Kauffman, L. H. (2013). Nothing but the truth: A short dialogue. *Cybernetics & Human Knowing, 20*(3-4), 188-190.

Deely, J. (2001). *Four ages of understanding: The first postmodern survey of philosophy from ancient times to the turn of the twenty-first century.* Toronto: University of Toronto Press.

Deely, J. (2012). *Purely objective reality* (Semiotics, Communication and Cognition: Vol. 4). The Hague: De Gruyter Mouton.

Misak, C. J. (1995). *Verificationism: Its history and prospects.* London: Routledge.

Peirce, C. S. (1982-). Writings of Charles S. Peirce: A chronological edition, (Peirce Edition Project, Eds.). Bloomington, IN: Indiana University Press. (8 volumes published so far. Cited as Wvolume:page)

Bunnell, P. (2007). *Flocking*. Un-retouched photograph.

Cybernetics and Human Knowing. Vol. 22 (2015), nos. 2-3, pp. 19-20

Remembering Ranulph Glanville

Mary Catherine Bateson[1]

As president of the American Society for Cybernetics, Ranulph Glanville found a context in which to explore his concept of design and his understanding of how organizations function. This led him to an increasingly inclusive understanding of the field of cybernetics/systems theory and to experimentation in designing programs for the annual meetings. His leadership followed the first generation, in which the concepts of cybernetics were shaped by the pioneers involved in the Macy conferences (1946-1953) and the era when computer science exploded, leaving systems theory somewhat in the shade. His death leaves the Society still struggling to define its place in the Academy as both a discipline and a bridge between disciplines. My contacts with him stemmed from the involvement of both my parents, Margaret Mead and Gregory Bateson, in the Macy conferences, and the different ways in which cybernetics influenced their continuing work, especially Mead's thinking about culture change and Bateson's thinking about learning, psychotherapy, and the environmental crisis.

Ranulph approached cybernetics from design, and when he became involved in second order cybernetics, he adopted a proposal of my mother's, that the Society attend to the *cybernetics of cybernetics*, that is, that it be self-conscious about the nature of its knowledge base and the processes of developing and passing it on. He has had a consistent interest in the history of the field and the personalities involved. Early on he adopted a comment I had made that "the time to learn cybernetics is in kindergarten," which I now recognize as an assertion that the basic concepts of cybernetics need to be incorporated in early learning so that they become *common sense,* rather than treated as rarified subjects for specialized expertise. Child development specialists have described infants learning about simple lineal causality and experimenting with it, discovering, for instance, that Momma would repeatedly pick up and return a toy thrown on the floor—but we still know very little about learning related to circular causation or attention to multiple causal factors and complex side-effects. Our politics and economics tend to be based on oversimplified ideas of causality and purpose, with potentially disastrous consequences.

As president of the Society, Ranulph persistently maintained a presence for and dialogue with the arts, which I have found particularly relevant to my own work in comparing the choices individuals make over a lifetime with the processes of artistic creation. At his request, I became a reader for his D.Sci (following two previous doctorates) and spent a week in a borrowed New York apartment plowing through an extraordinarily diverse and challenging range of publications. I did not then become a

---

1. Email: mcatbat@gmail.com

consistent member of the Society or participant in the annual meetings, but Ranulph was extremely persistent in drawing me back.

For the 2014 meeting of the ASC, Ranulph designed a discussion of the future of cybernetics but was unable to be present due to his illness. From discussions on the phone, I believe he was hoping for a major change of course, perhaps something that would radically change the status of cybernetics in the Academy and in Western Culture, from an immensely useful tool to a source of philosophical and ethical insight. Cybernetics and Buddhism are the two strands of discourse that may enable a shift from current emphases on competition and individualism to an emphasis on cooperation and interdependence that would not only make possible a coordinated effort to mitigate climate change but also propose paths to the resolution of international conflict. Ranulph Glanville seems to have had a prophetic understanding of a basic cultural shift of which we are sorely in need.

Bunnell, P. (2004). *Silt and Sand.* Un-retouched photograph.

Cybernetics and Human Knowing. Vol. 22 (2015), nos. 2-3, pp. 21-25

Ranulph Glanville's Thesis on the Theory of Objects and the Invention of Second-order Cybernetics

Albert Müller[1]

An unambiguous history of second-order cybernetics is still to be written despite the fact that its fundamentals are quite well known (see Scott, 2011; Müller, 2008a, 2008b; Müller & Müller, 2011; Glanville, 2009, 2012a, 2014). The ways its origin or invention may be described depends on which of its dimensions are preferred and given a paramount role. This might be first, the category of the observer as seen as part of what could be defined as an observed system; or it might be the category of self-referentiality, self-reflexivity or self-application. If the category of the observer takes the foreground we find early origins of second-order cybernetics already in Ashby's (1956) *An Introduction to Cybernetics* or in Pask's (1961) *An Approach to Cybernetics*. Into the center of interest the category of the observer has been moved by Maturana's (1970) famous quote "everything said is said by an observer." At about the same time Heinz von Foerster also dealt with the category of the observer until he ended up with a final definition of second-order cybernetics: "First-order cybernetics, the cybernetics of observed systems; Second-order cybernetics, the cybernetics of observing systems" (von Foerster 1974, p. 1).

The other dimension of second-order cybernetics, involving self-referentiality, self-reflexivity, and self-application, was for the first time stressed by social-anthropologist Margaret Mead (1968). In a key-note paper given on the occasion of the first annual conference of the American Society of Cybernetics (Foester et al., 1968) she suggested that the problems a formal society for cybernetics would face should be treated and solved using cybernetics itself. So her paper published with the title "Cybernetics of Cybernetics" stands for a quite practical example of self-application. Heinz von Foerster later told the story that the title was actually his idea since at the time he acted as the editor of the conference publication. This account seems to be reliable but we should keep in mind that Foerster's (and others) actual development of the ideas and practice of second-order cybernetics took some more years.

At about the same time and mostly independently during the first years of the 1970s Ranulph Glanville worked on comparable problems. Glanville, then a student at the architecture school of the Architectural Association, met Gordon Pask in 1966 (according to Gordon Pask's diaries). This meeting immediately triggered his interest in cybernetics. After his AA diploma Glanville started working on a PhD in

---

1. Email: albert.mueller@univie.ac.at

Cybernetics at Brunel University (supervised by Gordon Pask) with highly original results. Glanville's 1975 thesis is entitled *The Objects of Objects (The Points of Points) Or Something About Things. (Also Known As: A Cybernetic Development On Epistemology and Observation. Applied to Objects in Space and Time)*. A full version of the dissertation never appeared in print in full length; a variation of a part of it was included in the first volume of the *Black B∞x* series (Glanville 2012). The dissertation has been quoted in quite a number of articles by Glanville, but very rarely elsewhere, apparently due to the limited numbers of available copies. The theory of Objects which was the main topic of Glanville's PhD work was taken up by him afterwards as well in some of his articles and lectures (most of them now included in Glanville, 2012a). Some of these were also translated into German by Dirk Baecker for a book by Glanville, *Objekte* (1988). *Objekte* contains translations of eleven articles from 1978 to 1987 and an introduction from 1988 offering an impressive overview over Glanville's thought and work from the late 1970s to the late 1980s. During his lifetime Glanville dealt with a large number of other topics but returned to the theory of Objects when he gave his Viennese Heinz von Foerster-Lecture titled "Grounding Difference" in 2004 (see Glanville, 2007). At the annual conference of the ASC in Bolton 2013 Glanville offered a tutorial on his theory of Objects.

Even though he cites the relevant literature (and in such a way demonstrating that he was scientifically up-to-date in the field) Glanville's dissertation emerged mostly independently from other scientific activities which brought second-order cybernetics into being. In form and in content Glanville's work resembles to some extent as much a piece of art as a scientific text. The design of it does not comply with the requirements of a classic academic thesis as an ordinary student would write it. Regarding its form there are clear precedents at least for the main part of the dissertation. The first one is Ludwig Wittgenstein's *Tractatus Logico-Philosophicus* which is very well known for being divided into seven parts, six of them being divided according to a hierarchical numbering system. Glanville makes use of a similar numbering system in order to structure the definitions and propositions which constitute the main part of his dissertation. But it is not only Wittgenstein's *Tractatus* which worked as an inspirational source, there is also one article by Heinz von Foerster (1972/74) "Notes on the Epistemology of Living Things." In this article, von Foerster offered twelve propositions which he commented on in a separate section of the article (named "notes"). This is exactly the pattern Ranulph Glanville made use of in his thesis. I am not able to decide whether Glanville was inspired by von Foerster's article or whether he arrived at this organizational principle on his own. Glanville does reference von Foerster's work.

I mentioned the unconventional and innovative dimension of Glanville's thesis which makes it a work of art. This is not only made clear by having chosen a quote from Samuel Beckett as a motto for the dissertation and by referencing Beckett in the acknowledgements, it is also made clear by the fact that a part of the appendix consists of "stories" whose literary quality is undisputed although their didactic character should not be missed. Another part of the appendix consists of the scores for a piece of music. Glanville's affinity to Beckett goes back to his youth. As a sixteen-year-old young man he wrote to Samuel Beckett—and received an answer addressing "Mr. Ranulph Glanville" from the pen of the 40 years older person who later won a Nobel prize. Glanville's interest in music maybe started even earlier. His stepfather, who also was his instructor, was a gifted musician and composer.

The formal structure of the unpublished thesis is the following:

Abstract
Corrections
Dedication
Thesis Motto
Preface
Acknowledgements
Contents
Reference Marks and Numbering Systems Used In The Text
(Excluding Technical Terms in the Glossary) Terminology
Introduction
Main Text
Explanatory Text
Experiment Reports
    Experiment One: Report: (London Knowledge Test)
    Experiment Two: Report (London Structure Test)
    Experiment Three: Report (Conceptual Space)
    Experiment Four: Report (Leadenhall Market)
Conclusions
Stories
    When You First See This Place
    Now That You Can See Me
    Seeing Is Believing Oblique Killing
    A Fred Blogg's (R.I.B.A.) Eye View
    A Superman's Eye View
    A God's Eye View
Appendices
    Appendix l: Correspondence
    Appendix 2: Signification in Frege's Triangle
    Appendix 3: Tune into Memories of You
Glossary of Terms Used
Bibliography

The structure and form of the thesis clearly refers to its contents which, of course, cannot be discussed here sufficiently and therefore adequately. Some characteristics should, however, be mentioned. A first and central feature of Glanville's thesis consists of a completely new version of the concept of Object (nota bene: Glanville did write *Object* with a capital letter, not *object*, to clearly show his break with tradition). Glanville rejects or overrides the distinction or separation between subject and object which has been a key position of Western philosophy. The term *subject* is completely ignored in the main part of the thesis (there is one exception on p. 79). In Glanville the concept of Object becomes radical and unconventional. His Objects acquire completely new features, they observe themselves and other objects, and they are (inter-)actors. (Many years later science studies theorists like Bruno Latour or Hans-Jörg Rheinberger were to get recognition and popularity for similar or comparable ideas.)

One of the key elements of Glanville's radicalizing of the definition of the Object is to ascribe to it the attribute of self-observation and self-referentiality. An Object which would not observe itself would be no Object ("For an Object to exist in our Universe, it must be observable" [Glanville, 1975, 0.1, p. 15]). In a more conventional view, an object could be just a stone lying "there" merely determined by the laws of gravity and nothing else. Glanville's Objects follow a completely different maxim, they observe and they (inter-)act.

In this context Glanville also speaks of the double role of the Object as observing and being observed. Starting from this distinction (observing/observed) Glanville creates a new (pictorial) syntax in order to represent logical structures. One of the advantages of Glanville's sign-system lies in the fact that logical operations, that is, something quite abstract (like Boolean operations), are immediately visualized. This is another point where the author turns out to be not only a philosopher but a designer first and foremost.

Following the theoretical (main) part of the dissertation with its notes and comments, Glanville adds experiments and case studies he carried out over the years before writing down the thesis either alone or with the help of students and colleagues (who are all mentioned by name). All of the case studies were inspired by basic questions of—and Glanville's highly innovative conception of—the problems of design, architecture and urban planning with which he had been familiar since his training at the Architectural Association. Experiments and case studies, by Glanville et al., could have been carried out by social scientists or social psychologists, as well. But Glanville was able to provide convincing interpretations which tie the experiments to the main theoretical part of the thesis.

Portions of literary texts, and music scores, also relate to the theoretical part of the thesis, but it is left to the reader to build a bridge between these chapters.

Ranulph Glanville (2012a, p. 193) once described his work in the following way: Cybernetics often is be described as a meta-science; second-order cybernetics therefore would constitute a meta-meta-science, and so his own work on the theory of Objects would be part of a meta-meta-meta-science (see also Scott, 2011, p. 407). Such a description seems to me—at least in relevant parts—misleading when applied to his dissertation. In my view, Ranulph Glanville's theory of Objects as presented in his thesis of February 1975 represents a genuine contribution to second-order cybernetics and, equally, to radical constructivism (many years later Glanville [2012b] stressed precisely such a kind of equation). As I see it, Glanville's thesis is a radicalized version of second-order cybernetics (and not something meta meta meta), a version touching territories the much older (co-)creators of second order cybernetics did not penetrate in the first half of the 1970s. The double aspect of the Object as observing and being observed, is what he theorized; or even more the quadruple aspect of self-observing, externally observing, self-observed, externally observed Objects. Taking this double/quadruple aspect into account, the Foersterian distinction between first and second-order cybernetics is somehow canceled by extending and reformulating it significantly. Glanville's Objects, because of their abilities for self-observation, take their place in the second order and cannot be removed from there.

References

Ashby, W. R. (1956). *An introduction to cybernetics.* London: Chapman & Hall.

Glanville, R. (1975). *The objects of objects (the points of points) or something about things.* (Also known as: *A cybernetic development on epistemology and observation. Applied to objects in space and time.*) Unpublished Thesis. Brunel University, London.

Glanville, R. (1988) *Objekte.* (D. Baecker, Ed. & Trans.). Berlin: Merve.

Glanville, R. (2007). Grounding difference. In A. Müller & K. H. Müller (Eds), *An unfinished revolution? Heinz von Foerster and the Biological Computer Laboratory | BCL 1958-1976* (pp. 361-406). Vienna: Echoraum.

Glanville, R. (2009). *The black box: Vol. 3. 39 Steps.* Vienna: Echoraum.

Glanville, R. (2012a), *The black box: Vol. 1. Cybernetic circles.* Vienna: Echoraum.

Glanville, R. (2012b). Radical constructivism = second-order cybernetics. *Cybernetics & Human Knowing, 19*(4), 27-42.

Glanville, R. (2014). *The black box: Vol. 2. Living in cybernetic circles.* Vienna: Echoraum.

Maturana, H. R. (1970). Neurophysiology of cognition. In P. Garvin (Ed.), *Cognition: A multiple view* (pp. 3-23). New York: Spartan Books.

Mead, M. (1968). Cybernetics of cybernetics. In H. von Foerster, L. J. Peterson, & J. K. Russel (Eds.), *Purposive systems: Proceedings of the first annual symposium of the American Society of Cybernetics* (pp. 1-11). New York: Spartan Books.

Müller, A. (2008a). Zur Geschichte der Kybernetik: Ein Zwischenstand. *Österreichische Zeitschrift für Geschichtswissenschaften, 19*(4), 6-27.

Müller, A. (2008b). Computing a reality. Heinz von Foerster's lecture at the A.U.M. Conference in 1973. *Constructivist Foundations, 4*(1), 62-69.

Müller, A. & Müller, K. H. (Eds.). (2007). *An unfinished Revolution? Heinz von Foerster and the Biological Computer Laboratory | BCL, 1958-1976.* Vienna: echoraum.

Müller, A. & Müller, K. H. (2011). Systeme beobachten. Über Unterschiede und Gemeinsamkeiten von Kybernetik zweiter Ordnung und Konstruktivismus. In B. Pörksen (Ed.), *Schlüsselwerke des Konstruktivismus* (pp. 564-582). Wiesbaden, Germany: VS Verlag für Sozialwissenschaften.

Pask, G. (1961). *An approach to cybernetics.* London: Hutchinson.

Scott, B. (2011) *Explorations in second-order cybernetics. Reflections on cybernetics, psychology and education.* Vienna: Echoraum.

Von Foerster, H. (1974). Notes on the epistemology of living things. In H. von Foerster (2003), *Understanding understanding: Essays on cybernetics and cognition* (pp. 247-259). New York: Springer. (Originally published in 1972)

Von Foerster, H. (Ed.). (1974). *Cybernetics of cybernetics.* Urbana: University of Illinois.

Von Foerster, H. (2003). *Understanding understanding. Essays on cybernetics and cognition.* New York: Springer.

Von Foerster, H., White, J. D., Peterson, L. J., & Russel, J. K. (Eds.) (1968). *Purposive systems: Proceedings of the first annual symposium of the American Society of Cybernetics.* New York: Spartan Books.

Wittgenstein, L. (1922). *Tractatus logico-philosophicus.* London: Kegan Paul.

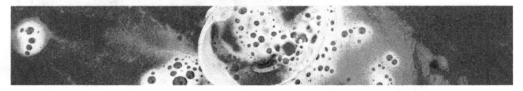

Bunnell, P. (2013). *Between Waves.* Un-retouched photograph.

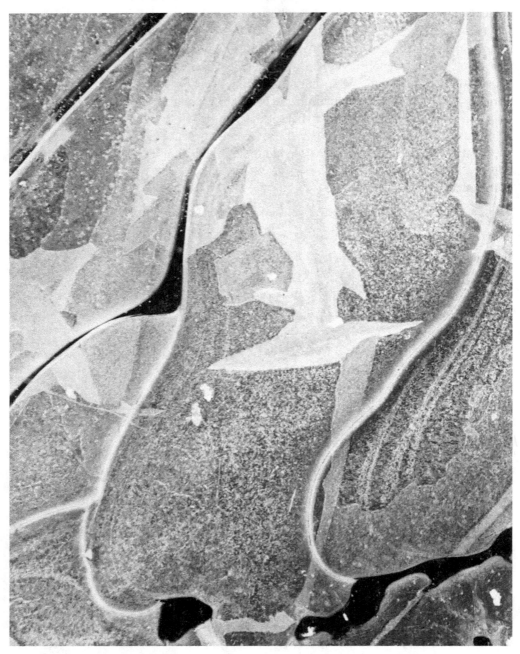

Bunnell, P. (2010). *Sinuous Solid*. Un-retouched photograph.

Cybernetics and Human Knowing. Vol. 22 (2015), nos. 2-3, pp. 27-47

De Profundis: Ranulph Glanville's Transcendental Framework for Second-order Cybernetics[1]

Karl H. Müller[2]

Ranulph Glanville was a prolific writer, a magic designer, an avant-garde musician, a cybernetician of the first- and of the second-order, a philosopher in disguise, to name only a few roles. His contributions to second-order cybernetics and to areas like design, philosophy, conversation theory, methodology or games, with the tools and perspectives of his version of second-order cybernetics were collected under the title *The Black B∞x* in three volumes in edition echoraum (Glanville, 2009, 2012, 2014) and were ordered and arranged by Ranulph Glanville himself so that they allow a general and systematic overview on this very large, diverse, and impressive corpus.

In this short essay I will undertake a systematic attempt to make this work more easily accessible for others, including myself, and to provide a special location for Ranulph Glanville within the research program of second-order cybernetics in particular and within the research tradition of radical constructivism in general. It will become my central thesis in this article that Ranulph Glanville's special role and function was to provide a meta-approach to all the available research programs in radical constructivism. This framework was transcendental in nature and focused on the conditions of the possibility for observation, for communication, for language, for knowledge or for learning to emerge at all. Thus, Ranulph Glanville reserved a unique place for himself that, at the same time, turned out to be magic for his explorations and very difficult to grasp for his intellectual environment.

Keywords: epistemology, transcendental approaches, objects, second-order cybernetics, conversation theory, design

"De profundis" is the title of psalm 130 which is used in the Christian liturgical tradition for the faithfully departed or within the Jewish culture for the High Holidays. "De profundis" was also the title of a letter by Oscar Wilde towards the end of his time in prison which he wrote to Lord Alfred Douglas, his former companion, and in which Oscar Wilde found his road to salvation in Jesus Christ in the unusual role of a romantic artist. De profundis is necessarily bound and tied with the fundamentals of life and death, and with the available roads for a life worthwhile to pursue. Thus, de profundis deals with last things. I chose this psalm as the title of this article because Ranulph Glanville played persistently on fundamental grounds and he departed just a while ago with a strong faith in the power and the sustainability of second-order cybernetics especially in the years to come.

---

1. I want to dedicate this article to Aartje Hulstein, who accompanied and supported Ranulph for decades and I hope that she finds my remarks on Ranulph relevant. And I wish to thank Alexander Riegler and Bernard Scott for their very helpful comments on an earlier version of this article.
2. Steinbeis Transfer-Center New Cybernetics, A-1160 Vienna. Email: office@wisdom.at

Ranulph Glanville's contributions to second-order cybernetics, as collected in the three volumes of *The Black B∞x* in edition echoraum (2009, 2012, 2014), stretch over many areas and disciplines and are, even at second sight, not easily understood in their levels of abstraction and not particularly well-known within the scientific community.

Thus, my main goal for this article is the construction of a small Glanville-map which covers the main domains in the three volumes of *The Black B∞x* and which provides a basic orientation for an interested reader. Aside from my systematic attempt to make this work more easily accessible, my second goal is to find the special location for Ranulph Glanville's framework within the research tradition of radical constructivism.

In this small article it will become my central thesis that Ranulph Glanville's special role and function was to provide, by necessity, the most general and the most profound framework for radical constructivism because he dealt with the necessary pre-conditions for communication, for language, for conversation or for control to emerge at all. He offered to us a theoretical background of what we usually take implicitly for granted when we engage in research activities.

I will start my journey to Ranulph Glanville's oeuvre with the question of why his work was not recognized as highly relevant during his lifetime. In particular, why was Ranulph Glanville, unlike Heinz von Foerster, Humberto R. Maturana or Francisco J. Varela, not a central part of the Hall of Fame for radical constructivism?

1. Unequal Fame Distribution in Radical Constructivism

The Hall of Fame for radical constructivism was invented and built mostly for the German speaking world around Siegen in Germany by Siegfried J. Schmidt as the responsible architect and cognitive planner in the formative years during the 1980s. This special Hall of Fame, while an important milestone for radical constructivism, exhibited also a small number of strange effects. The first peculiar feature of the Hall of Fame of radical constructivism was that many official members in this Hall were strongly opposed to being categorized as radical constructivists. Heinz von Foerster, for example, expressed his strong mental reservations against the term radical constructivism by saying: "I don't want to be associated with constructivism. I have no idea what constructivism means" (2001, p. 241; Translation by K.H.M.). Likewise, Humberto R. Maturana and Francisco J. Varela were clearly opposed to being classified as radical constructivists. Despite these strong mental reservations the term *radical constructivism* and the corresponding Hall of Fame followed an astonishing career path.[3]

- Ernst von Glasersfeld actually invented the term *radical constructivism* in 1974 which he reserved initially for Jean Piaget only. But Ernst von Glasersfeld used this concept quite frequently for others, including himself, as well. Through his

---

3. On the varieties of radical constructivism, see also Riegler (2015). And on the astonishing career paths of radical constructivisms from 1974 onwards, see also Müller (2008a, 2008b, 2010, 2011).

self-propagations and through his publications, lectures, workshops, conferences and media reports an initial drift towards the term radial constructivism set in already prior to 1980. Due to the initiative of Ernst von Glasersfeld, even an autobiography of radical constructivism was being fabricated with Heinz von Foerster as co-author under the title "How we invent ourselves" (Foerster & Glasersfeld, 1999). Moreover, the radical constructivists as seen by Ernst von Glasersfeld were also strongly linked in a personal network since the 1960s. Thus, Ernst von Glasersfeld's frequent and persistent use of the term *radical constructivism* from 1974 onwards prepared the terrain for a new and apparently homogeneous research tradition.

- In 1987 Siegfried J. Schmidt edited a book under the title *The Discourse of Radical Constructivism* (Schmidt, 1987) which in retrospect constitutes the basic design for the Hall of Fame of radical constructivism since the 1980s. This volume comprised articles by Humberto R. Maturana, Francisco J. Varela, Heinz von Foerster, and Ernst von Glasersfeld as well as a group of German authors who concentrated their work on special aspects of radical constructivism. In this founding document of radical constructivism S. J. Schmidt wrote a long introductory article on radical constructivism as a new paradigm for inter- or transdisciplinary discourse. In this overview, radical constructivism appears as a coherent research tradition which was jointly generated by the works of Humberto R. Maturana (1970) and Francisco J. Varela (1979) in the domain of biology or by Heinz von Foerster and Ernst von Glasersfeld in the area of cognition and philosophy.

- Another important step in the diffusion of radical constructivism was the publication of three volumes with articles by Humberto R. Maturana (1985), Heinz von Foerster (1985) and Ernst von Glasersfeld (1987). These three books were published in German again under the guidance cf Siegfried J. Schmidt who became the dominant network organizer for radical constructivism during these years.

- Already in the year 1984 Niklas Luhmann published his *Social Systems* (Luhmann, 1984) which contained very frequent references to the later stars of radical constructivism like Heinz von Foerster, Humberto R. Maturana and Francisco Varela plus related persons like George Spencer-Brown or W. Ross Ashby. This raised the interest of the sociological or the wider social science community in these authors and contributed, in turn, to the inclusion of Niklas Luhmann in the Hall of Fame itself. By the early 1990s *radical constructivism* gradually emerged in the German speaking parts of the world as a common label for a coherent research tradition and a post-modern world-view.

- Nevertheless, the German Hall of Fame for radical constructivism had its hidden biases and its open favorites which become more transparent if one looks at the bibliography of Siegfried J. Schmidt's overview on radical constructivism where he cites most frequently a group of German authors (Hejl [6], Roth [6] and Schmidt [9]), followed by radical constructivists like Heinz von Foerster [2],

Ernst von Glasersfeld [4], Humberto R. Maturana [2] and Francisco J. Varela [3]. But the management of the Hall of Fame of radical constructivism was unwilling or unable to incorporate other groups of authors as well. For example, Jean Piaget who was the reason for Ernst von Glasersfeld's invention of the term radical constructivism in the first place, plays only a marginal role and is cited only once. Additionally, it becomes important to recognize that the British tradition of first- and second-order cybernetics, headed by W. Ross Ashby, Stafford Beer (1972, 1974, 1975, 1979), Gordon Pask (1975a,1975b, 1976) or Grey Walter as well as by Gordon Pask's collaborators like Ranulph Glanville, Bernard Scott or Paul Pangaro was seemingly placed at the outside of the inside of radical constructivism.

- Furthermore, the German Hall of Fame of radical constructivism showed also a very unequal distribution for its shares of fame, which apparently followed a power-law distribution. A power-law distribution implies that a very small number of scientists receive a very large portion of fame whereas most researchers receive no shares of fame at all.

- Finally, the German Hall of Fame, due to its overall design, had to include a central room which, however, was usually locked for the general public and was not used by the group of established radical constructivists. No hints could be found in the open exhibitions by radical constructivists that this room in the Hall of Fame must be accessible even by sheer necessity. Until today this central room looks quite abandoned. For strange reasons, the contributions and exhibits of Ranulph Glanville are stored in this central room. And Ranulph Glanville, up to this day, is still the only author in this special place.

This short story was intended to emphasize that Ranulph Glanville was excluded twice from the German Hall of Fame of radical constructivism. On the one hand, Ranulph Glanville was not given a proper space for public viewing like the von-Foerster-Hall, the Humberto R. Maturana-Hall, the Francisco J. Varela-Hall, and so forth, and, on the other hand, the big names in this Hall of Fame, with the exception of Niklas Luhmann, did not quote, refer to, or mention Ranulph Glanville for particular insights, unique arguments or new concepts, even under the most obvious circumstances.[4]

Aside from this dual exclusion or, alternatively, dual closure, Ranulph Glanville was not easy to grasp for colleagues, friends or students, although he was exceptionally open, friendly and very interested in what others were doing. But Ranulph Glanville was specialized in being non-specialized. Due to his multiple talents in cybernetics, architecture, design, music, science writing, poetry or philosophy he offered probably too much to his intellectual peers who had a much

---

4. As a clear example I will only mention Heinz von Foerster's article "Objects: Token for Eigen-behaviors" from 1976 (von Foerster, 2003) where he cites only Warren McCulloch and Jean Piaget. But Heinz von Foerster was Ranulph's PhD examiner in 1975 when Ranulph presented his PhD thesis with the unofficial title *The Objects of Objects, the Points of Points, or—Something about Things*.

more restricted repertoire of skills and qualifications and, thus, a comparatively limited potential for understanding.

At this point I want to undertake the challenging task of providing an overview of the general approach of Ranulph Glanville and its constitutive elements. In a series of approximation steps I would like to come close to the basic organization of his overall theoretical and methodological perspective. I would like to make his work more accessible for others, including of course myself.

2. The Organization of Research Programs and Research Traditions

Initially, several core concepts for this article, namely the terms *research program*, *research tradition, radical constructivist network* or *second-order cybernetics* require a few additional comments because they are to be used in a rather specific way.

Starting with the concept of a research program, Figure 1 shows the typical modules or building blocks for empirical research programs like a theoretical core (TC), a set of methods, models and mechanisms (MMM), linked to the theoretical core, the embeddedness of TC and MMM within a wider background-knowledge BK, the bridge-modules BM which link the theoretical domain with applications, a class of paradigmatic examples A_1 to A_n, that is, applications of TC, MMM and BM on observable or actually observed processes as well as an underlying class of observations, data and measurements (DT).[5]

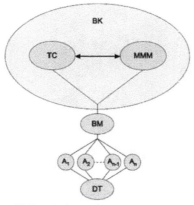

BK: Background Knowledge   TC: Theoretical Core MMM: Models, Mechanisms, Methods BM: Bridge Modules
A: Applications   DT: Data and Measurements

Figure 1: Mapping Empirical Research Programs

In Figure 1, no arrows have been used in order to stress the duality of top-down and bottom up flows. Theoretical concepts, generative mechanisms or transfer-

---

5.   As a relevant selection from the philosophy of science literature, see Balzer, Moulines, and Sneed (1987), Curd & Cover (1998), or Stegmüller, Balzer, and Spohn (1981).

modules are as much shaped by the DT-segment as observations, methods and data are determined by the theoretical core, the MMM-segment or the BM-domain.

Radical constructivism will be introduced as a research tradition despite the self-propagating history of the radical constructivist movement. Research traditions can be described as a network of research programs and can be visualized in the way of Figure 2.

Here one can see networks of theoretical cores T, of methods, mechanisms and models, of bridge modules M on the one hand and a rich network of different classes of observations, methods and data (DT) and a network of wider application domains (D) on the other hand. The application area changes into larger application domains D1 to Dn where each domain captures a set of paradigmatic examples.

Radical constructivism, due to its large set of application domains and to its heterogeneous composition of theories, generative mechanisms or models, can best be characterized as a trans-disciplinary research tradition. The application domains D_i cover unusually wide areas, ranging from cell-biology, to organizations and societal evolution.[6]

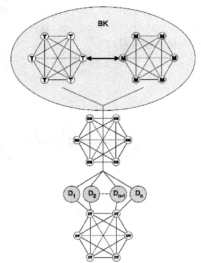

BK: Background Knowledge T: Network of Theoretical Core M: Network of Models, Mechanisms, Methods
BM: Network of Bridge Modules   D: Application Domains   DT: Network of Data and Measurements

Figure 2: Mapping Empirical Research Traditions as Networks of Research Programs

In an organizational perspective, the concept of a radical constructivist network stands for a group of visionary scientists[7] who developed the cognitive building

---

6. It should be emphasized at this point that radical constructivism as a research tradition is limited to small groups in biology, the cognitive sciences or the social sciences in which observers play a central role. Karin Knorr-Cetina (1989), for example, distinguishes between social constructivism of the Berger & Luckmann (1966) variety, epistemological constructivism and empirical constructivism which covers the program of laboratory studies. Radical constructivism as understood by me covers the epistemological branch only.

blocks for radical constructivism from the 1960s and 1970s onwards. The radical constructivist-network during its formative years in the 1960s and 1970s includes, in alphabetical order, W. Ross Ashby, Stafford Beer, Heinz von Foerster, Ranulph Glanville, Ernst von Glasersfeld, Humberto R. Maturana, Gordon Pask, Jean Piaget, Ricardo Uribe, Francisco J. Varela and others and grew in the 1980s and 1990s by adding Dirk Baecker, Søren Brier, Hans-Rudi Fischer, Peter Hejl, Klaus Krippendorff, Niklas Luhmann, Paul Pangaro, Gerhard Roth, Siegfried J. Schmidt, Bernard Scott, Fritz B. Simon, Stuart A. Umpleby, Paul Watzlawick, and so forth. Second-order cybernetics will be used throughout this article as a special research program within the radical constructivist research tradition. Second-order cybernetics was promoted especially by Heinz von Foerster from the late 1960s onwards and constitutes, thus, an important component within radical constructivism as a research tradition. According to Ranulph Glanville, Heinz von Foerster's contributions to the formation of second-order cybernetics lie in three areas. First, Heinz von Foerster realized, following Margaret Mead's important lecture at the founding conference of the American Society for Cybernetics (ASC) (Mead, 1968) that cybernetics, like so many other scientific disciplines and fields, can be applied to itself. The second important element was Heinz von Foerster's emphasis on observers and their inclusion in research processes as a methodological revolution in science. Finally, the third component is linked to the second one and affirmed that our ways of world-making are actively constructed and, thus, observer-dependent.[8]

Thus, second-order cybernetics which, as emphasized by Ranulph Glanville (Glanville, 2012, p. 175), was developed in the short period between 1968 and 1975 can be characterized as a research program with building blocks like application of understanding to self (Glanville, p. 193), ethics, stability from within, self-reference, conversational communication and striving toward "improvement, not perfection" (Glanville, p. 194).

Likewise, autopoiesis stands for a special research program under the radical constructivist umbrella. Autopoiesis was created by Humberto R. Maturana in collaboration with Francisco J. Varela and Ricardo Uribe from the 1970s onwards and forms, thus, another significant element within the radical constructivist research tradition.[9]

Within this broad research tradition of radical constructivism and within second-order cybernetics as research program Ranulph Glanville occupied, as will be shown below, a special and singular location and function.

---

7. On the importance of visionary leaders, see, e.g., Hollingsworth and Hollingsworth (2000, 2011).
8. On the research program of the BCL in general and on Heinz von Foerster in particular, see Müller (2007).
9. The same description can be provided, in principle, for Gordon Pask and his conversation theory, for Niklas Luhmann's theory of systems or for Stafford Beer's management cybernetics. All three research programs can be considered as network components of the radical constructivist research tradition.

3. Ranulph Glanville's Transcendental Framework: 3 Initial Approximations

In my view, the major difficulty for others to assess the content of Ranulph Glanville's work has to do with the unusual scope and the goals which Ranulph Glanville set for himself to accomplish. These goals for his endeavor claimed a unique position for himself which can be summarized as becoming a constructor of last resort. Like a lender of last resort who offers support where no one else is capable of doing so, Ranulph Glanville wanted to provide the necessary backgrounds, whatever others produced in the foreground.

Thus, as my first approximation step to Ranulph Glanville, I would like to summarize the goals of his general framework. He viewed himself in a position which offers the necessary generalizations and abstractions of what his intellectual in-group within radical constructivism or within second-order cybernetics produced.

My work might be thought of as a generalization of the work of others. (Glanville, 2012, p. 192)

In another version, Ranulph Glanville saw himself in the role of the principal designer who produces the possibilities for others to pursue their work.

Many of us watch games. Some play, others umpire or referee. Still others govern games and make/ remake the rules. There are some who create the field of play, mark and maintain it. The potential (Gibson's affordance) is not however, limited to games: others may use the ground in a completely unanticipated way, unintended by those who set up the ground. Behind all these is the person who creates the possibility of the blank field on which all this potential can be expected. That person is me, and that is my work: I create the unformed, empty field. (Glanville, 2012, p. 35)

Still in another variation, Ranulph Glanville placed his work at the level n+1 under the assumption that the level n was the highest and most general level explored by the rest of radical constructivists.

Cybernetics is often considered a meta-field. The Cybernetics of Cybernetics is, thus, a meta-meta-field. My work is, therefore, a meta-meta-meta-field. (Glanville, 2012, p. 192)

Aside from being most general by necessity, the central research question for Ranulph Glanville was phrased by him in the following manner:

My major initial concern was to develop a set of concepts that might explain how, while we all observe and know differently, we behave as if we were observing the same 'thing'. What structure might support this? (Glanville, 2012, p. 192)

Thus, the crucial research problem for Ranulph Glanville was transcendental in nature because he was searching for the conditions of the possibility for observing, knowing, communicating, and so forth and operated, therefore, on a very special level of abstraction.

Due to this unusual abstraction level Ranulph Glanville abolished also a separation which has become almost trivial and self-evident in the course of the long-term evolution of science, namely the separation between scientific problems on the one hand and problems in artistic or in applied domains like education or architecture on the other hand. In the case of Ranulph Glanville we are dealing with a general framework which is so much broader or wider than the normal scope of conventional research programs, which covers very heterogeneous territories and which crosses easily the borders and boundaries between basic science, applied science, design, arts or technology.

A second approximation step to the transcendental framework of Ranulph Glanville can be reached by looking carefully on the overall organization which he used for ordering his contributions to second-order cybernetics in the very comprehensive edition of his *Black Box* (Glanville, 2009, 2012, 2014), especially for volume 1 and volume 2. Volume 3 under the sub-title *39 Steps* contains a collection of his cybernetic musings in the Journal *Cybernetics and Human Knowing* which has become the current platform for discussing the frontiers and the potential of second-order cybernetics, autopoiesis and cyber-semiotics (Brier, 2008, 2009).

Volume 1 with the title *Cybernetic Circles* contains five major sections, namely cybernetics, objects, black box, distinction and variety, volume 2 has the title *Living in Cybernetic Circles* and is composed of five domains, namely design, representation, knowing, education, and others where others is focused mainly on tributes to Gordon Pask, Heinz von Foerster, Gerard de Zeeuw, Alfred Locker, Ernst von Glasersfeld and Richard Jung.

From the differentiation in the two volumes one could infer that Ranulph Glanville established the theoretical foundations of his transcendental framework in volume 1 and applied this new framework to areas like design or education in the second volume. But such an inference would miss the special form in which Ranulph Glanville organized his theoretical or analytical production on the one hand and his practical work on the other hand. In fact, the sub-title of volume 2 *Living in Cybernetic Circles* also applies with equal strength to volume 1 as well. Whatever he did, Ranulph Glanville arranged it in a permanent circle of theory and practice or doing and reflecting, following Donald A. Schön's concept of a reflective practitioner.[10] Thus, this action-reflection circle can be described as a circular organization between the context of production where a problem solution is reached, an artistic object is created or a technological system is constructed on the one hand and the context of reflection on the other hand which deals with the consequences, including ethical issues, further implications or modifications of the finished product, future developments, etc. Ranulph Glanville lived constantly within these production-reflection circles and shifted positions frequently, taking circularity seriously and as a stable form for theoretical or practical action.[11]

---

10. Donald A. Schön defined reflection in a circular manner as an action that leads to new actions (Schön, 1983)
11. It is worthwhile to mention Paulo Freire, who emphasized that though these action-reflection circles we are able to develop a critical consciousness (Freire, 1973).

As my third approximation step I want to describe the special role which Ranulph Glanville reserved for himself within these cybernetic circles. Due to his involvement in second-order cybernetics the answer seems trivial, because second-order cybernetics is, after all, the cybernetics of observing systems. But much was or is written about second-order cybernetics or observers where the authors of these texts remain in the shadow or implicit. Here, Ranulph Glanville differed radically from most of the usual observer-inclusive accounts.

In order to demonstrate the difference which makes a difference I want to propose two new terms, namely the concepts of an endo-mode and of an exo-mode which are both possible ways of world-making or of exploring the world. The distinction between an endo-mode and an exo-mode can be traced back also to Heinz von Foerster who developed a very intriguing list of characteristic differences between two fundamentally different epistemic attitudes towards one's world or environment.[12]

> Am I an observer who stands outside and looks in as God-Heinz or am I part of the world, a fellow player, a fellow being? (Foerster, 2014, p. 128)[13]

Subsequently, Heinz von Foerster provides us with the following list of distinctions which can be used directly for the differentiation between an endo-mode and an exo-mode.

Table 1: Dichotomies for the Exo-Mode and for the Endo-Mode

Exo-Mode	Endo-Mode
Appearance	Function
World and I: separated	World and I: one
Schizoid	Homonoid
Monological	Dialogical
Denotative	Connotative
Describing	Creating
You say how it is	It is how you say it
Cogito, ergo sum	Cogito, ergo sumus

12. It must be mentioned that Otto E. Rössler published a book on endo-physics (1992) which raised considerable interest. See, e.g., Atmanspacher and Dalenoort (1994). However, the distinction developed here between an exo-mode and an endo-mode differs significantly from the exo- and endo-differentiation by Otto E. Rössler, who assumes a two-level structure of reality.

13. For Heinz von Foerster, the decision between an exo-mode from without or an endo-mode from within belongs typically to the undecidable questions whose charm it is that they have to be decided by us.

This shift to the I of the observer or to the endo-mode requires profound methodological changes because, following Heinz von Foerster once again, "I am the observed relation between myself and observing myself" (Foerster, 2003, p. 257) and *I* invites a host of additional notions like self-reference, self-observing, self-reflexivity, and the like plus a long history of paradoxes, based on the notions of I and self.

It concludes my third approximation step to Ranulph Glanville's framework that he operated fully in and with the circularities of this endo-mode which included himself in explicit form. For decades, Ranulph Glanville worked as a pioneer who explored the very strong ramifications of an observing from within or of the endo-mode in defining concepts, in developing models, in designing, in learning interactions, etc. Whatever he wrote, his co-presence as an author or observer was never implicit, but always explicit.

4 The General Transcendental Framework of Ranulph Glanville

Turning to the theoretical elements it will come as minimal surprise that Ranulph Glanville's transcendental framework is nowhere to be found in the work of other radical constructivists, including myself.

As pointed out already, the level of operation can and must be considered as the most general, the most abstract or the most fundamental because he deals with the presuppositions with or with the conditions of the possibilities for language, communication and knowledge or, in a Glanville-like fashion, with language of language, communication of communication and knowledge of knowledge. Thus, my fourth approximation step to Ranulph Glanville's work refers to the building blocks with which he operated.

The initial and starting point for Ranulph Glanville's approach lies in his notion of Objects with capital *O*. Objects are not to be confused with what we usually view as objects, like a cup of tea, a pair of socks or books on a shelf. Objects can be used with many different connotations.

> Think of an object of study; objecting to someone's argument; the objective ...; an object in action ...The word also inverted its meaning over time; originally, object meant what subject now means. I liked this ambiguity ... I have more recently begun to use the expression "object of attention" to help explain what the Objects are: finding this expression took me about 20 years! (Glanville, 2012, p. 40)

Thus, in a proto-ontological environment Objects can be considered as the elementary and irreducible units which, in principle, can be humans, machines, animals, robots, aliens, observers or other active creatures. Ranulph Glanville offers a very illuminating distinction between Objects and things in the usual meaning of the word.

> Our observing is not of: it is. If we insist it is of, then it is of observing. We do not observe things. We observe observing. If we insist there should be 'things' to be observed, these 'things' come about through our constructing. (Glanville, 2012, p. 397)

Objects are "structural forms" and "are structurally determined by the inclusion of the observer, with the difference between each observer preserved" (Glanville, 2012, p. 228). Thus, Objects and observing as the mode of operation become the building blocks in Ranulph Glanville's participatory universe (see also, Scott, 2012).

As my fifth approximation step to Ranulph Glanville's transcendental framework, Objects, due to the proto-ontological stage, cannot be characterized by attributes and relations, but they have to be glued together by at least one primordial principle. Here, Ranulph Glanville develops one fundamental principle which is postulated like an axiom and is regarded as self-evident.

Glanville's fundamental maxim is the principle of mutual reciprocity. This principle requires a "reciprocal arrangement, by which what may be of one may be of the other" (Glanville, 2012, p. 205) For Objects or systems alike this principle demands that "when a quality is attributed to one system, there must be a potential for the same quality to be attributed to the system it is distinguished from" (Glanville, p. 194) The principle of mutual reciprocity has far-reaching implications for Objects and the universe of observing "which is, of course, the universe of Second Order Cybernetics" (Glanville, p. 40).[14]

> To enter into a universe of observation, I must observe myself. If I substitute the word know for observe, the form of the answer is the same: I know myself. I generalised this formulation to all inhabitants of the universe (of observing): they must all be assumed to observe themselves. There is no way to test this; it is simply a condition for entry into the universe – an assumption or axiom of this system of understanding. (Glanville, 2012, pp. 40-41)

Thus, Objects play dual roles of self-observing and self-observed and the switch between these two roles generates, following Ranulph Glanville, time. This invocation of time "permits a unique Object to both observe and be observed – in sequence – without infringing the quality of uniqueness." (Glanville, 2012, p. 327)

From the reciprocal distinctions of self-observing and observing others the basic configuration in the universe of Objects can be summarized as follows:

$$<O> = [(X)P] \rightarrow E$$

where P and E "are the observing and observed roles of Objects" (Glanville, 2012, p. 324), X stands for the "Modelling Facility" or for the role potential of Objects and

> O indicates the whole Object (the 'I'), the triangular brackets < > indicate the naming of the whole Object ..., the equal sign = indicates the idea 'consists in', the square brackets [ ] contain the act of observation, the round brackets ( ) contain that through which the action takes place, and the arrow → indicates something like 'gives rise to'. (Glanville, 2012, p. 324)

---

14. Ranulph Glanville regarded "the moment of insight, that to join a universe of observing I must observe myself, as the greatest moment of insight and intellectual joy that I have ever been fortunate enough to experience" (Glanville, 2012, p. 41).

In the universe of Objects role switching between observing and being observed at different times requires a special action "for the continuing existence of a cognizant self" (Glanville, 2012, p. 321) which, for Ranulph Glanville, becomes Memory. Memory produces itself, constructs the identity of Objects and guarantees that Objects can observe themselves and others differently. Thus, observing, role-switching, time and Memory become the initial configuration for Ranulph Glanville's Objects which, due to the principle of mutual reciprocity, can be found throughout this universe of observation.

Moreover, the work of George Spencer-Brown played a very important function in the development of the theory of Objects because Ranulph Glanville used it as testing and experimental ground for his own transcendental framework. The section "Distinction" in Glanville (2012), can be seen as a critical appraisal of Spencer-Brown's *Laws of Form*. As a short-cut, Ranulph Glanville showed in various ways that Spencer-Brown's first command of drawing a distinction cannot, "as it were, bring things into being" (Glanville, 2012, p. 470) because drawing a distinction requires certain conditions which, thus, leads to the question "how do these conditions come into being?" (Glanville, p. 470) For a reader, it could be a worthwhile exercise to follow the arguments more closely how Ranulph Glanville, together with Francisco J. Varela, attempts to escape the infinite regress inherent in this distinction-conundrum (Glanville & Varela, 2012).

In sum, Ranulph Glanville's theory of observing Objects, including the primordial principle of mutual reciprocity, offers a most general playground for everybody and everything and provides the necessary foundations of second-order cybernetics.

> The Theory of Objects is, in my opinion, the central missing element in the essential formulations that make up second order cybernetics, a sort of glue that holds the other essentials together and without which second-order cybernetics would remain somewhat detached, incoherent and inarticulate. Objects accommodate the centrality of observing, and the significance of the observer in this universe, as well as delimiting the unknowable and the impossible. (Glanville, 2012, p. 41)

Additionally, the theory of Objects provides the necessary conceptual basis for differences and autonomy between these observing units.

> Objects allow us to think as we will/do. They are containers for freedom, for individual difference in e.g., meaning, action and understanding. They recognize that in observing we are necessarily participants, no matter how withdrawn: that to think is to act, for any concept we form, we form by acting. (Glanville, 2012, p. 41)

The sixth approximation to Ranulph Glanville's framework lies in the specification of his paradigmatic heuristic model which, not surprisingly, is the Black Box in the version of W. Ross Ashby. The Black Box becomes "the universal model for how we come to construct our understanding through which we name the world" (Glanville, 2012, p. 42) The Black Box is "perhaps the most astonishing intellectual device we have yet produced, above even, Occam's Razor" (Glanville, p. 42) in exploring the universe of objects. At the same time, the Black Box demonstrates the fundamental ignorance or an "essential obscurity" (Glanville, p. 179) for finding patterns or explaining the behavior of objects because black boxes are "essentially and crucially a construct of the observer" (Glanville, p. 179). They allow "us to find patterns of behaviour that are based on ignorance and remain grounded in that ignorance" (Glanville, p. 42). The patterns which can be generated with the Black Box only exist

> as an explanation created by the experimenter: an explanation that is of a non-existent box, with no inside or outside, imagined into place by the experimenter. What's in it can't be examined because there is no box! (Glanville, 2012, p. 42)

Finally, my seventh step in the approximation to Ranulph Glanville's transcendental framework concerns the re-formulation and re-interpretation of several core concepts of traditional cybernetics, especially the notions of control, goals and variety. In Ranulph Glanville's participatory and reciprocal universe the usual separations between a controlling unit and a domain under control do no longer hold, because the controller is at the same time under the control of the controlled. "Control is circular, not linear. It exists between the controlling and the controlled systems" (Glanville, 2012, p. 523). Likewise, goals have to be analyzed as a circular interplay between the goals of the system under observation and the goals of an observer of a particular system, variety becomes an attribute of the system as well as a measure of an observer, exploring the variety of a system, and so forth.

As an important corollary, Ranulph Glanville dismisses the universal validity of W. Ross Ashby's law of requisite variety, which postulates that for effective control the states of the controlling system must be equal or greater than the states of the controlled system. Within his second-order or endo-universe, variety can increase in time, but the varieties of controlling and controlled systems must be equal.

> The variety of each (controlling/controlled) system in a control circularity, as determined by an observer, must be identical. (Glanville, 2009, p. 121)

Thus, Ranulph Glanville presents a very coherent transcendental account of second-order cybernetics in which participation and circularities offer a stimulating most general playground for all possible games, scientific and otherwise. With these seven approximation steps Ranulph Glanville's overall perspectives and goals become, so I hope, better understandable for a reader. Obviously, each of these seven approximation steps is described in much more detail in volume 1 of the three volume set of the *Black B∞x* and was represented here as a preliminary intellectual Glanville-map for the high theoretical territories.

5   Adding Bridges into Special Domains

In addition to the general transcendental framework Ranulph Glanville offers additional specifications which can be used in special areas like management, design, education, architecture or communication. Ranulph Glanville sees the relations between the first and the second volume of the *Black Box* in the following way.

> Volume 2 focuses on the sorts of understandings developed in volume 1 as they play out in other areas: their implications and what they have to offer these areas ... Volume 2 shows how these understandings throw light, explain and can effect changes in the performance of various areas. (Glanville, 2014, p. 11)

These articles can be seen as bridge-modules as depicted in Figures 1 and 2 which, however, are not detached or separate from the general transcendental framework. They reflect it and are based on it. For the reader these articles exhibit fully the paradoxes of the hermeneutic circle. One has to understand the general transcendental framework in order to grasp the implications in special areas and one has to fully comprehend these specifics in order to understand the general transcendental framework.

For these special domains[15] I will use a recombination of the areas in volume 2 and another list which Ranulph Glanville used for the praxis of second-order cybernetics in general (Glanville, 2012, p. 194ff.)[16] Here I want to draw the attention of the reader to some highlights and will omit other also relevant contributions.

Language, Communication and Society
After specifying the structures that support the necessary differences in observing Objects, Ranulph Glanville needed an answer to the second part of his initial research question why we behave as if we were observing the same 'thing'. With support from Gordon Pask who was Ranulph Glanville's teacher and mentor and who developed an impressive general framework under the name of *conversation theory*, based on recursive interactions (Pask, 1975a,1975b, 1976), Ranulph Glanville also found an answer to the second puzzling issue in which he elaborates on the basic requirements for conversations to occur and to succeed. Within Pask's framework, Ranulph Glanville contributed a series of requirements for conversations to occur at all.

First, he stressed the importance of three levels which should be accessible within a conversation, namely "the level of the conversation, the meta-level, in which monitoring and error regulation takes place, and the substratum, which provides the context." (Glanville, 2009, p. 101)

Second, he listed a series of requirements for a successful conversation which he classified as operational requirements and as inspirational requirements which bind the participants in a conversation in an equal and a symmetrical fashion. The operational requirements include routines like comparing, monitoring, error-correction, etc. and are based on the different understandings of a topic by the participants in a conversation. Among the inspirational requirements one finds

---

15. The articles in volume 3 can be viewed as special pieces which are self-contained and combine elements from the general transcendental framework with actual problems.

16. Two domains in the list of the praxis of second-order cybernetics are left out, namely computation and mathematics on the one hand and management on the other hand because they were not central domains for his explorations.

necessary attributes like respect for others and their differences, openness, a readiness to change one's own position, to admit errors, and so forth.

These requirements become also interesting from a media-theoretic and societal point of view because they are mostly absent in the world of modern media and their ways of orchestrating talks or discussions which turn out more closely related to noise than to conversations.

It would also be a worthwhile research task to compare the Glanville requirements for conversations with the ideal speech situation of Jürgen Habermas, especially because Jürgen Habermas' work (1984a,1984b) is not mentioned by Ranulph Glanville, neither in the text nor in the bibliography.

Learning, Education, Knowledge and Cognition

Within these four broad categories I want to draw the reader's attention to one specific element, namely to knowledge. The first point is rather obvious and it can be phrased as an extension of Humberto R. Maturana's theorem number one: "Anything said is said by an observer" (Foerster, 2003, p. 283):

Anything known is known by an observer.

Ranulph Glanville provides a short, but compelling argument for this claim.

> There is nothing known without a knower. Or rather, there is nothing we can know without us knowing it. It is possible that something might be known without a knower ..., but we could not know this. (Glanville, 2014, p. 459)

The second point is linked with a heuristic device which Ranulph Glanville uses very frequently and that is the rich diversity in meanings and contexts which we usually overlook. Thus, he emphasizes over and over that knowledge appears in a variety of forms like in embrained, embodied, encultured, embedded or encoded knowledge (Glanville, 2014, p. 484). Moreover, knowledge may come in an implicit manner within skills and routines which have no encoded basis and an important aspect of knowledge creation lies in the transformation from an implicit to an explicit status.

In sum, Ranulph Glanville is able to generate a comprehensive and densely populated game-field for knowledge analyses in which varieties of knowledge types, different knowing observers and the authors of these knowledge studies, their goals and their methods become entangled in a fascinating way. Moreover, popular notions like knowledge-societies or knowledge-based societies become, following the analyses by Ranulph Glanville, far from obvious and clear because they can mean, in principle, so many different and even contradicting features, trends or characteristics.

Design

Living in his cybernetic circles, Ranulph Glanville established also a circle between cybernetics and design with "design as cybernetics in practice, cybernetics as design in theory." (Glanville, 2014, p. 12)

I have … developed the analogy between second-order Cybernetics and design so as to give mutual reinforcement to both. Design is the action; second-order Cybernetics is the explanation." (Glanville, 2012, p. 200)

For him, design offers, aside from the usual contexts of this concept, a special meaning in relation to innovations and novelty. For Ranulph Glanville, design becomes a special routine or process in which something new for an observer is created.

Design, as interpreted here, is a (the) process of making the new (Glanville, 2012, p. 200).

Moreover, design becomes, due to its novelty creation, also important for research and for research methodologies.

Design is the key to research. Research has to be designed. Considering design carefully … can reveal how better to act, to do research—to design research … But design should be studied in design's terms. For, design is the form, the basis. And research is a design act." (Glanville, 2014, p. 164)

Finally, Ranulph Glanville establishes an intriguing connection between complexity and design which becomes highly critical of the current discourse on the new brave complexity world where complexity enters as a ubiquitous and universal explanatory variable. Within the endo-mode in general and within Ranulph Glanville's participatory universe observers "choose the complexity (they) see in this world" (Glanville, 2014, p. 293).

Along these circles, the journey to the coherent manifold of contributions by Ranulph Glanville must come to an end. With this short mapping of the transcendental Glanville-landscapes his perennial existential problems become, I hope, more intelligible to the reader. Ranulph Glanville was working and publishing throughout his entire life outside of philosophy departments and outside the philosophic community which would be the only places and audiences left in the Western world where one can pursue transcendental issues. And he was constantly talking to persons from cybernetics, systems science, architecture, design, and so forth for whom new transcendental problem solutions fell clearly beyond their expertise and also beyond their immediate interests. Despite a different impression at first sight, Ranulph Glanville pursued a coherent, but enormous and enormously ambitious project throughout his life within rather unsuitable and partly counter-productive intellectual environments with clearly different agendas. In this sense, Ranulph Glanville became the tragic or romantic hero of second-order cybernetics.

6   Varieties of Foundational Configurations

As my final point near the end of this article I want to locate the particular place of Ranulph Glanville's Object-oriented framework in the very long history of philosophy in general and of metaphysics and epistemology in particular. With his primary focus on Objects, the primordial reciprocity principle, transcendental issues and cybernetic tools Ranulph Glanville produced a highly original contribution to contemporary metaphysics or epistemology, albeit without much resonance in philosophical circles or in scientific settings, including cybernetics itself.

With respect to different metaphysical or epistemological perspectives on primordial configurations and foundational problems I can see basically six broad traditions in the long history of philosophy with different initial starting points, different primordial configurations and, thus, different foundations.

- Configuration I—the world "as it is": The first and most powerful tradition has its focus on the world or the environment and demands of sense-driven observers to adapt or accommodate to these external constraints and conditions. Hypothetical realism or empiricism have become the main vehicles for this particular configuration for which David Hume in his *An Enquiry Concerning Human Understanding* (1748) can be named as the most prominent and most consequential proponent.
- Configuration II—the composition of the world: The second approach asks and searches for the most fundamental components of the world or environment, including observers themselves. In the history of philosophy, Georg Friedrich Leibniz offered an advanced version of a compositional foundational framework in his "Monadology" in the year 1714 with monads "without windows" as the basic simple substances.
- Configuration III—the power of rational observers: The third foundational tradition is concentrated on the cognitive organization of observers as knowledge producers. For the third approach, Immanuel Kant in his three *Critiques*, published between 1781 and 1790, the later variations within German idealism or various versions of phenomenology can be seen as the most influential groups within this tradition.[17]
- Configuration IV—communicative practices of observers: The fourth foundational configuration places its attention on practices or actions of observers as the basic foundational units. Karl Marx with his inversion of Hegel's perspective or, in recent times, Ludwig Wittgenstein in his *Philosophical Investigations* (1953) or George Spencer-Brown in *Laws of Form* (1969) have become important milestones in this foundational tradition.[18]
- Configuration V—the logic of language: The fifth configuration is relatively recent in origin, can be characterized by its linguistic turn and develops, according to Michael Dummett (2014) or Richard Rorty (1967), from Gottlob Frege's work on arithmetic and especially from the early writings of Ludwig Wittgenstein's

---

17. It should be added that in recent years Siegfried J. Schmidt (2003, 2010) developed a research program within this tradition which is based on *Setzungen* (settings)
18. An early version can be found in Johann Wolfgang von Goethe's *Faust* in the phrase: *Und schreib getrost, im Anfang war die Tat.*

Tractatus logico-philosophicus (1921) where the logic of language becomes the primary foundational focus.

* Configuration VI—relational organizations: A sixth foundational configuration sees a system of relations or formal structures as its primary focus for further explorations. Relational structures as foundations (Kineman, 2011) can be found in quantum mechanics, for example in the works of John A. Wheeler (1994), Lee Smolin (1997) or Carlo Rovelli (1997), in space-time analyses (Kauffman, 1994, Schmeikal 2010, 2012, 2014), in biology (Rosen, 1991), or in philosophy (Kaipayil, 2009).

Ranulph Glanville's *Tractatus logico-objectivus* appears as a cybernetic version within the second tradition which is based on a universe of observing Objects. Of course, the choice between these six foundational traditions belongs to the core of undecidable problems which have to be decided by us.

However, an interesting observation can be made that could lead to the decision that all six foundational frameworks are necessarily incomplete. In fact, whenever one starts with foundational problems, one has to begin with elements from all of the six configurations simultaneously: world or environment, units of composition, the cognitive organization of observers, practices of observers, language and relations. Omit language, and the problems cannot be posed at all. Omit practices, and nothing can happen. Omit the cognitive organization, and the results become incomprehensible. Omit the basic units of composition and no one will pose the problem. And, finally, omit relations and no structures can emerge. As in the case of Foerster's unity of cognition thesis,[19] all six elements are needed simultaneously, namely a network of compositional units with highly developed cognitive architectures and a rich action potential, including language games within a world or environment. This foundational Eigen-form which reproduces itself again and again once foundational problems are raised becomes, thus, the necessary starting and fixed point for observing, communicating or knowing to happen.

7   De Profundis, Once Again

From the deep and with deep sorrow we have to cope with the loss of probably the most fundamental and most elementary author within the research tradition of radical constructivism. I hope that I succeeded in persuading the interested reader that Ranulph Glanville leaves a most profound and most general transcendental framework for second-order cybernetics which combines epistemology, design, ethics and cybernetics into, following Douglas R. Hofstadter, an eternal golden braid.

For me as friend I will miss the opportunities to initiate a conversation with Ranulph. During the last years my and probably our favorite places for conversations

---

19. See von Foerster (2003, p. 105) or Foerster and Müller (2003) where Heinz von Foerster speaks of cognitive processes like perceiving, remembering and inferring, and of the necessity to consider the totality of cognitive processes simultaneously.

were located in Orlando, Florida, where Nagib Callaos continues to organize a very social and very interactive type of world-conference on systemics, cybernetics and informatics each year. In the afternoon we used to sit in a bar along the swimming pool of the conference hotel and discussed profound and hot topics in the very warm and humid Florida climate with calypso-music in the background. For me, these Orlando dialogues with Ranulph came very close to what we can achieve, under the best of circumstances, as happy moments for human beings.

I view this article as a small contribution to a huge debt we owe to Ranulph Glanville and which we, friends, colleagues and students alike, were unable to return to him during his lifetime. To a large part, we missed the opportunities to play on the grounds which he generously opened and offered.

References

Atmanspacher, H., & Dalenoort, G. J. (Eds.). (1994). *Inside versus outside. Endo- and exo-Concepts of observation in physics, philosophy and cognitive science*. Berlin: Springer-Verlag

Balzer, W., Moulines, C. U., & Sneed, J. D. (1987). *An architectonic for science: The structuralist program*. Dordrecht: Reidel

Beer, S. (1972). *Brain of the firm: A development in management cybernetics*. New York: Herder and Herder.

Beer, S. (1974). *Designing freedom*. New York: Wiley

Beer, S. (1975). *Platform for change*. London: Wiley

Beer, S. (1979). *The heart of enterprise*. New York: Wiley

Berger, P. L., & Luckmann, T. (1966). *The social construction of reality: A treatise in the sociology of knowledge*. Garden City, NJ: Anchor Books

Brier, S. (2008). *Cybersemiotics: Why information is not enough*. Toronto: University of Toronto Press

Brier, S. (2009), Cybersemiotic pragmaticism and constructivism. *Constructivist Foundations, 5*(1), 19-39.

Curd, M., & Cover, J. A. (Eds.). (1998). *Philosophy of science: General issues*. New York: W.W. Norton

Dummett, M. (2014). *Origins of analytic philosophy*. London: Bloomsbury Academic.

Foerster, H. v. (Ed.). (1974). *Cybernetics of cybernetics*. Urbana: University of Illinois

Foerster, H. v. (1982). *Observing systems*. Seaside, CA: Intersystems Publications

Foerster, H. v. (1985). *Sicht und Einsicht. Versuche zu einer operativen Erkenntnistheorie*. Braunschweig, Germany: Friedrich Vieweg und Sohn

Foerster, H. v. (2001). Rück- und Vorschauen. In A. Müller, K. H. Müller, & F. Stadler (Eds.), *Konstruktivismus und Kognitionswissenschaft. Kulturelle Wurzeln und Ergebnisse. Heinz von Foerster gewidmet* (2nd ed., pp. 229-242). Vienna: Springer.

Foerster, H. v. (2003). *Understanding understanding: Essays on cybernetics and cognition*. New York: Springer

Foerster, H. v. (2014). *The beginning of heaven and earth has no name: Seven days with second-order cybernetics*. New York: Fordham University Press

Foerster, H. v., & Glasersfeld, E. v. (1999). *Wie wir uns erfinden. Eine Autobiografie des radikalen Konstruktivismus*. Heidelberg, Germany: Carl Auer Systeme Verlag.

Foerster, H. v., & Müller, K. H. (2003). Action without utility: An immodest proposal for the cognitive foundations of behavior. *Cybernetics and Human Knowing, 10*(3-4), 2-50.

Freire, P. (1973). *Education for critical consciousness*. New York: Seabury Press

Glanville, R. (2009). *The black b∞x: Vol. 3. 39 steps*. Vienna: edition echoraum

Glanville, R. (2012). *The black b∞x: Vol. 1. Cybernetic circles*. Vienna: edition echoraum

Glanville, R. (2014) *The black b∞x: Vol. 2. Living in cybernetic circles*. Vienna: edition echoraum

Glanville, R., & Varela, F. J. (2012). Your inside is out and your outside is in. In R. Glanville (Ed.), *The black b∞x: Vol. 1. Cybernetic circles* (pp. 469-478). Vienna: edition echoraum,

Glasersfeld, E. v. (1974). Piaget and the radical constructivist epistemology. In C. D. Smock, & E. von Glasersfeld (Eds.), Epistemology and education (pp. 1-24). Athens: Follow Through Publications.

Glasersfeld, E. v. (1987). *Wissen, Sprache und Wirklichkeit. Arbeiten zum radikalen Konstruktivismus*. Braunschweig, Germany: Friedrich Vieweg und Sohn.

Hollingsworth, R. J., & Hollingsworth, E. J. (2000). Radikale Innovationen und Forschungsorganisation. Eine Annäherung. *Österreichische Zeitschrift für Geschichtswisenschaften, 1*, 31 - 66

Hollingsworth, R. J., & Hollingsworth, E. J. (2011). *Major discoveries, creativity, and the dynamics of science*. Vienna: edition echoraum

Kaipayil, J. (2009). *Relationalism: A theory of being*. Bangalore, India: JIP Publications.

Kauffman, L. H. (1994). Space and time in computation, topology and discrete physics. In *Proceedings of the workshop on physics and computation. Physcomp '94* (pp. 44 - 53). New York: IEEE Press.

Kauffman, L. H. (2005). Eigen-forms. In *Heinz von Foerster in memoriam* [Special issue]. *Kybernetes, 34,* 129-150.

Kauffman, L. H. (2009). *Laws of form and the logic of non-duality.* Paper presented at the Conference on Science and Non-Duality, October 2009, San Rafael, CA.

Kineman, J. (2011). Relational science: A synthesis. *Axiomathes, 21*(3), 393-437

Knorr-Cetina, K. (1989). Spielarten des Konstruktivismus. *Soziale Welt, 40*(1/2), 86 -96.

Luhmann, N. (1984). *Soziale Systeme.* Frankfurt: Suhrkamp ·

Mead, M. (1968). Cybernetics of cybernetics. In H. von Foerster, L. J. Peterson, & J. K. Russel (Eds.), Purposive systems (pp. 1-11). New York: Spartan Books.

Maturana, H. R. (1970). *Biology of cognition* (Biological Computer Laboratory Research Report, BCL 9.0). Urbana, IL: University of Illinois

Maturana, H. R. (1985). *Erkennen. Die Organisation und Verkörperung von Wirklichkeit* (2nd ed.). Friedrich, Germany: Vieweg und Sohn.

Maturana, H. R., Varela, F. J. (1980). *Autopoiesis and cognition: The realization of the living.* Dordrecht: Reidel.

Maturana, H. R., Varela, F. J. (1987). *Der Baum der Erkenntnis. Die biologischen Wurzeln menschlichen Erkennens.* Munich: Scherz Verlag

Müller, K. H. (2007). The BCL—An unfinished revolution of an unfinished revolution. In: A. Müller, & K. H. Müller (Eds.), *An unfinished Revolution? Heinz von Foerster and the Biological Computer Laboratory | BCL, 1958-1976* (pp. 407-466). Vienna:edition echoraum.

Müller, K. H. (2008a). Non-dualistic? Radical constructive? *Constructivist Foundations, 3*(3), 181-191.

Müller, K. H. (2008b). Methodologizing radical constructivism. Recipes for RC-designs in the social sciences. *Constructivist Foundations, 3*(4), 50-61.

Müller, K. H. (2010). The radical constructivist movement and its network formations. *Constructivist Foundations, 6*(1), 31 - 39.

Müller, K. H. (2011a). The two epistemologies of Ernst von Glasersfeld. *Constructivist Foundations, 6*(2), 220 - 226

Müller, K. H. (2011b). Heinz von Foerster and the self-reflexive turn. *Cybernetics and Human Knowing, 18*(3-4), 133-138.

Müller, K. H., & Riegler, A. (2014). Second-order science: A vast and largely unexplored science frontier. *Constructivist Foundations, 10*(1), 7 – 15

Pask, G. (1975a). *The cybernetics of human learning and performance: A guide to theory and research.* London: Hutchinson Educational

Pask, G. (1975b). *Conversation, cognition and learning: A cybernetic theory and methodology.* New York: Elsevier

Pask, G. (1976). *Conversation theory: Applications in education and epistemology.* New York: Elsevier

Riegler A. (2015 in-press). What does the future hold for radical constructivism? In J. Raskin, S. K. Bridges, & J. S. Kahn (Eds.), Studies in meaning 5: Perturbing the status quo in constructivist psychology. New York: Pace University Press, New York.

Rössler, O. E. (1992). *Endophysics. Die Welt des inneren Beobachters.* Berlin: Merwe Verlag. (with a foreword by Peter Weibel)

Rorty, R. M. (Ed.). (1967). *The linguistic turn.* Chicago: Chicago University Press.

Rosen, R. (1991). *Life itself: A comprehensive inquiry into the nature, origin and fabrication of life.* New York: Columbia University Press.

Rovelli, C. (1996). Relational quantum mechanics. *International Journal of Theoretical Physics, 35,* 1637-1678

Schmeikal, B. (2010). *Primordial space: The metric case.* Hauppauge, NY: Nova Science Publishers.

Schmeikal, B. (2012). *Primordial Space: Point-free space and logic case.* Hauppauge, NY: Nova Science Publishers.

Schmeikal, B. (2014). *Deccay of motion: The anti-physics of space-time.* Hauppauge, NY: Nova Science Publishers

Schmidt, S. J. (Ed.). (1987). *Der Diskurs des Konstruktivismus.* Frankfurt: Suhrkamp

Schmidt, S. J. (2003). *Geschichten und Diskurse. Abschied vom Konstruktivismus.* Reinbek, Germany: Rowohlt.

Schmidt, S. J. (2010). *Die Endgültigkeit der Vorläufigkeit. Prozessualität als Argumentationsstrategie.* Weilerswist, Germany:Velbrück Wissenschaft

Schön, D. A. (1983). *The reflective practitioner: How professionals think in action.* London: Temple Smith

Scott, B. (2012). Ranulph Glanville's Objekte. In R. Glanville (Ed.), *The black box: Vol. 1. Cybernetic circles* (pp. 63-76). Vienna: edition echoraum.

Smolin, L. (1997). *The life of the cosmos.* Oxford, UK: Oxford University Press.

Spencer-Brown, G. (1969). *Laws of form.* London: Allen & Unwin.

Stegmüller, W., Balzer, W., & Spohn, W. (1981). *Philosophy of economics.* Berlin: Springer

Varela, F. J. (1979). *Principles of biological autonomy.* New York: North Holland.

Varela, F. J., Maturana, H. R., & Uribe, R. (1974). Autopoiesis: The organization of living systems, its characterization and a model. *Biosystems, 5,* 187-196.

Wheeler, J. A. (1994). *At home in the universe.* New York: American Institute of Physics

Wittgenstein, L. (1921). Logisch-philosophische Abhandlung. In W. Ostwald (Ed.), *Annalen der Naturphilosophie,14,* 185–262

Wittgenstein, L. (1953). *Philosophical investigations.* Oxford, UK: Basil Blackwell.

Zeleny, M. (Ed.). (1980). *Autopoiesis: A theory of the living organizations.* New York: Elsevier.

Bunnell, P. (2004). *Sand Trees*. Un-retouched photograph.

Cybernetics and Human Knowing. Vol. 22 (2015), nos. 2-3, pp. 49-58

The Be-ing of Objects

Dirk Baecker[1]

The paper is a reading of Martin Heidegger's Contributions to Philosophy (Of the Even) by means of Ranulph Glanville's notions of black box, cybernetic control and objects as well as by George Spencer-Brown's notion of form and Fritz Heider's notion of medium. In fact, as Heidegger was among those who emphasized systems thinking as the epitome of modern thinking, did in his lecture on Schelling's *Treatise on the Essence of Human Freedom* a most thorough reading of this thinking, and considered cybernetics the very fulfilment of modern science it is interesting to know whether second-order cybernetics, as it was not known to Heidegger and as it delves into an understanding of inevitable complexity and foundational ignorance, falls within that verdict mere modernity or goes beyond it. If modern science in its rational understanding considers its subjects to be objects sitting still while being observed, then indeed second-order cybernetics is different. It looks into the observer's interactions with black boxes, radically uncertain of where to expect operations of a self, but certain that we cannot restrict it to human consciousness.

Ranulph Glanville was one of the most incisive epistemologists in cybernetics. He combined a thorough reading of first-order cybernetics concepts on black boxes, feedback, and circular control (Glanville, 1979, 1987; see Baecker, 1989) with a deep understanding of second-order cybernetics ideas on observers, distinctions, and productive because reflexive ignorance (Glanville, 1980, 1981, 1982). He called himself a radical constructivist (Glanville, 1996; see von Glasersfeld, 1995), but as a matter of fact he was more of an operational constructivist (Luhmann, 2002a, 2002b, 1997) in that he set out from ideas about design and came back to study how his constructions designed him (Glanville, 1999). Modelling yourself you discover how the other models you (Glanville, 1991).

In fact, a comprehensive reading of Glanville's papers places him firmly in an epistemological tradition that radicalizes the concept of the observer (Wiener, 1936), never flinches from attempting to learn to perceive (McCulloch, 2004), closely watches negations inherent in selections of explanatory constraints (Bateson, 2000a), and considers objects to be just another example of eigen-values stemming from recursive operations (von Foerster, 2003a, 2003b, 2003c). Indeed, nothing is more important than paying relentless attention to the distinction between ideas, notions, and notations (McCulloch, 2004) or between world, cognition and description (von Foerster, 1971).

Yet Glanville's writing and lectures may also be seen to follow up ideas pursued by older philosophical traditions. His own reflections on philosophy lead him back to Ludwig Wittgenstein, who in *Tractatus Logico Philosophicus* recommends treating one's own observations as traces on a screen leaving everything behind a mystery (Wittgenstein, 2001; Glanville, 1996). Light can also be thrown on Glanville's

---

1. Email: dirk.baecker@zu.de

thinking by reference to Plato's *Sophist,* who deeply mistrusts yet nevertheless relies on negations, contradictions, distinctions, and other kinds of polemic not only to verify but also to produce plausible explanations (Plato, 1993). I also wonder how Johann Gottlieb Fichte would have welcomed Glanville's modelling facility with regard to both observers and objects when, in his various takes on a *Wissenschaftslehre*, a "doctrine of scientific knowledge," he seeks to comprehend any object X and any identity $A = A$ as dependent on, and thus as elusive as an I positing *itself* (Fichte, 1970, 1986).

II.

I mention these grand themes only to prepare the ground for an even grander one. I suggest it might be fruitful to read some of Martin Heidegger's works within the perspective of both first-order and second-order cybernetics in general and Glanville's take on them in particular. It is well known that, when inquiring into the distinction of man from animal and stone, Heidegger was well acquainted with Jakob von Uexküll's *Theoretical Biology,* which may be seen as a close precursor to any thinking about observing systems (Heidegger, 1995; see von Uexküll, 1926). It may be less well known that Heidegger considered the notion of system, doing a close reading of it, to be the climax of a modern thinking intent not only on bringing everything, including knowledge, consciousness and subjectivity into the form of objectivity but to also on "systematizing" it according to different degrees and domains of being (Heidegger, 1985). This brings cybernetics to the fore as the kind of foundational science that encompasses all other sciences and thereby establishes man as the planner of a world, including the use of the arts as just another sort of vehicle for information (Heidegger, 1977). Operational research opts for and accepts only the light of reason, whereas for Heidegger the task of thinking consists in first indicating the clearance (*Lichtung*) concealed in any presence. To think the hidden within the plain may amount to thinking of complexity as not actually haunting but protecting any understanding from overemphasizing itself.

This is not the place to undertake a detailed reading of Heidegger's works with respect to systems thinking and cybernetics. I restrict myself to sketching out how Heidegger's notes in his *Contributions to Philosophy (Of the Event)* may relate to an epistemology of cybernetics as set out in Glanville's collection of his papers in *The Black B∞x* (Heidegger, 2012; Glanville, 2009-2015; see also Glanville, 1988; regarding cybernetics and Heidegger see also Baecker, in press; 1994). Assessments of Heidegger's *Contributions* vary strongly between considering it his second masterpiece after *Being and Time* or dismissing it as just a collection of redundant notes approaching an altogether enigmatic "turn" (*Kehre*) in his thinking, talking as it were about "a last God" able "to pass by" only if man is prepared "to go down" (see as an introduction to the literature Thomä, 1990: 761ff; Pöggeler, 1997; Polt, 2006; Mejía & Schüßler, 2009). Those notes—written in 1936–1938 when Heidegger was struggling to understand and free himself from his leanings towards German National

Socialism—were, under the conditions of his will, to be published only after the publication of his lectures of those years, as they provided the necessary background to the *Contributions*. In fact, the notes were called *Contributions to Philosophy* and subtitled *Of the Event* only in preparation for a book never written or at least never published, the planned title of which was *The Event*.

My reading of the *Contributions* is that they are a very thorough reconstruction and deconstruction of occidental metaphysics from Plato and Aristotle to Descartes, Leibniz, Kant and Nietzsche as a philosophy of the ontological difference between being and entity, focusing on the presence and objectivity of entities to eliminate, as it were, any magical or mystical thinking about hidden grounds, yet forgetting about being nevertheless being distinct from entity, reducing it to the most general and empty notion and thereby preparing the ground for installing first God and then the absolute as an a priori inaccessible to further inquiry.

Yet, I am not interested here in occidental metaphysics, even if I think important Heidegger's urge to understand ontological difference, the notion of system, and the "colossal machinations" of mankind in terms of a metaphysics which somehow conceals its own operations. What I am interested in is, indeed, Heidegger's turn to "another beginning" of philosophy, which reminds me of second-order cybernetics and of the laws of form as spelled out by George Spencer-Brown and somehow hesitatingly received by Glanville, here working together with Francisco J. Varela (Spencer-Brown, 2008; Glanville & Varela, 1981). Of course, this impression of mine may be attributed to my preconceptions and prejudices in reading Heidegger as well as towards Heidegger himself. As I see it, the distinction between figure and ground advanced by Gestalt theory (Wertheimer, 1925)—possibly supported by Marx's criticism of ideology (Marx & Engels, 1976), Freud's interpretation of dreams (Freud, 1996) and, even earlier, the novel's display of hidden motives and a liberal philosophy's understanding of words as distinct from thoughts (Locke, 1959, vol. 2, pp. 8ff; therefore looking at action to infer interests "which cannot lie"; Gunn, 1968)—prepared the ground for a new configuration of theory in the first half of the 20th century, which learned to look at the distinction of presence from absence, of the manifest from the latent, of the visible from the invisible, or of the symbol from the symbolized, and began to spell out the one in terms of the other. This new thinking found its perhaps most important expression in the discovery by both cybernetics and the mathematical theory of communication that any information is an understanding of a selection of a message from some technically determinate (Shannon & Weaver, 1963) or some physically, biologically or socially indeterminate, and so contingently constructed, set of possible messages (Baecker, 2013a). Heidegger's famous turn, though well prepared by his understanding of a being-there-toward-death (*Dasein zum Tode*), is another example of this new thinking in that he does away with any metaphysics that restricts itself to presenting objectivity and inquires into the time-play-space (*Zeit-Spiel-Raum*) yielding and conceding anything present in the first place.

III.

I think it is possible to describe Heidegger's *Kehre* as a turn to concepts that bear a striking resemblance to those of second-order cybernetics. The wording is different, but the underlying questions are close. In terms of Spencer-Brown's calculus, these questions are (i) how marked states are to be indicated and distinguished from unmarked states, (ii) who draws such a distinction, (iii) how the unmarked, being distinct from the marked, may in-form the marked, and (iv) how these operations of indication, distinction and re-entry bring forth a time and a space supporting and subverting those distinctions.

The scheme Heidegger conceives is that of an event, *E*, which has to be both worked out and received as an event linking an open world and a locked earth by man reckoning on gods to understand and undertake his being-there (Heidegger, 2012, No. 190):

$$\text{Man} \quad \left(\begin{array}{c} \text{World} \\ \uparrow \\ \leftarrow \text{E} \rightarrow \\ \downarrow \\ \text{Earth} \end{array} \right) \quad \text{Gods} \qquad \text{(There)}$$

To get there Heidegger develops a formalism similar to Spencer-Brown's. He speaks of a "joint" and "joining" (*Fuge* and *Fügung*) of thinking, consisting of an exercise reminiscent of certain Yoga practices as well as of Spencer-Brown's forms taken out of the form. The task is to mate (i) the appeal (*Anklang*) of a be-ing (*Seyn*) beneath and beyond the machinations of the present and colossal, (ii) the pass (*Zuspiel*) of a metaphysics relegating that be-ing to a general and empty notion of transcendent *idea*, *causa prima* or *condition of possibility*, (iii) the leap (*Sprung*) into the time-play-space where man has to ground his being-there and (iv) the return (*Rückkehr*) from the thrown-ness (*Loswurf*) of man designing himself as a future man attentive to the hint of the last god (Heidegger, 2012, No. 39 and 263).

How are we to understand this? Glanville's cybernetics begins and ends with the acknowledgment that any object is a black box. Everything of any interest for us, be it our life, our brain, our society, or the egg on our breakfast table, is concealed from us in terms of operations, functions and materiality (Ashby, 1958). We need assumptions about input and output to be able to observe an interaction with or among these objects. These assumptions are our assumptions and these observations are our observations, slowly but surely turning us, too, into black boxes as soon as we inquire about where these assumptions come from and how they can be ours. Yet, the interaction works, more often than not, thanks to ontogenesis and co-evolution, and produces white boxes inside which, to be sure, there are at least two black boxes still trying to get out (Glanville, 1982).

Thus, there is a closure, the black box, Heidegger's *earth*, and an opening, Heidegger's *world*, which is the interaction informing our observations without ever disposing of this foundational blackness. There is even a last God whose invitation (appeal) we accept when going beyond any present objectivity to ask for a truth that pervades the space where black boxes turn into white boxes and back again into black ones, and which will never assume any certainty but instead bears witness to the polemics (*aletheia*) inherent within the time-play-space of objects (Heidegger, 1998).

Glanville's interaction is Heidegger's leap. Both ground our being-there within a complex that is not to be reduced to either subjectivity or objectivity (see also Nietzsche, 2003, 154f, Notebook 9, autumn 1887, fragment No. 65). Instead, it is an opening and a closure at the same time. And there is nothing mystic about this since what we are dealing with is a form oscillating in itself (Kauffman, 1987). This is how we encounter a be-ing, which is indeed event and oscillation (*Gegenschwung*, see Heidegger, 2012, No. 133) and thereby reveals itself as a positively negating negativity (Heidegger, Nos. 144 & 267). Spencer-Brown's form provides us with an understanding of this peculiar operation of oscillation, because oscillation here means switching back and forth between two sides of a distinction that both negate and imply one another as in a paradox (Spencer-Brown, 2008, 45ff, 90ff & xf).

Heidegger's be-ing may be conceived of as Spencer-Brown's form giving play to an indication, a distinction, the operation of distinction, and the space brought forward by the operation. The operation is watched by an observer who is himself operating by means of distinctions. We deal with second-order observations to which all our terms relate while referring to objects with which we interact (Baecker, 2013). This, by the way, is why sociology, including and beyond any sociology of knowledge, is interested in the idea of distinction as an operation of second-order observers. There is something distinctly social in construction.

IV.

According to Heidegger there are two beginnings to occidental philosophy. The first we have behind us. It is called metaphysics and deals with the presence of objects backed by ideas, first reasons, God's creation, or transcendent conditions of possibility. Using Spencer-Brown's notation of the form of distinction we can write it as follows, clearly emphasizing entities at the cost of being whose state, not to speak of status, is more than uncertain even if remaining worth considering:

$$\text{object} = \overline{\text{entity}} \;\Big|\; \text{being}$$

The other beginning we have before us. It relates to the concern (*Geschichte* and *Geschick*) of a be-ing, which alone *is* whereas all entity *is not* (Heidegger 2012, No. 267). We can put it as follows, showing how an entity is both enriched and endangered by its distinction from be-ing being re-entered into the form of distinction:

$$\text{object} = \overline{\text{entity} \;\Big|\; \text{be-ing}}$$

Heidegger (2012) explains this form as giving precedence to the possible over the real. Thus any real entity, like a human preparing for death, is at any instant to be negated by the possible, while at any event this very possible brings itself forth again.

Glanville's objects do not account for this kind of negativity displaying itself. But Fritz Heider's things do (Heider, 1959). Heider's *things* are rigidly coupled elements, which may at any moment disintegrate into their *medium*, which consists of the same elements loosely coupled. While seeing things, we do not see the medium of light by which we are able to see. While listening to speech, we do not hear the medium of air bringing the sound of speech, together with noise, to our ears. Sociological theory has a similar understanding of the distribution media and success media of communication, which is in fact informed by Heider's understanding of media (Luhmann, 2012/13, vol. 1, chap. 2; see Parsons, 1977). We cannot go into this here. Suffice it to say that these media do indeed negate anything infixed in them by implying the possibility of other things. Heidegger warns us not to consider entity and be-ing, or thing and medium, to be immediately related to each other (Heidegger, 2012, No. 268). For Heider, media are *nothing* because we need things that allow us to infer the existence of the media by means of which we perceive these things (Heider, 1959, p. 13). We need theory to go back and forth between thing and medium as between figure and ground. And we need play to actually go for operations, be they perceptions, thoughts, actions or communication, which explore the time-space of a medium of things (Bateson, 2000b).

The negativity with which we are concerned is thus a thoroughly productive and therefore positive one. It negates any actual thing, object or entity by indicating a potential one. It calls on us to check for operations that make the potential real. Yet beware. The mood of this other beginning of occidental philosophy is not the wonder of the first beginning. Instead it is horror (Heidegger, 2012, No. 269). It is the horror of all observers, objects, entities and things displayed in a world in which they promptly decay. This is what Heidegger's event calls for and what second-order cybernetics eigen-values of recursive functions dispel for just a moment. And this is why for this other beginning Heidegger asks for time to take precedence over thought when accounting for both be-ing and being-there. Wonder allows you to engage in leisurely contemplation of the beauty of the cosmos while others do the work (Arendt, 1958). Horror instead lets you look at both sides of a distinction and at the contingency of the form they indicate.

V.

Two final remarks pointing to further issues. The first concerns one of the most pertinent questions in second-order cybernetics: how and where to posit a self. Metaphysics, of which first-order cybernetics and systems theory in its way is still a part admitted of a self only in the position of an observer, which is that of the subject looking at the world and accounting for its distinctions, notions, language, and reflection as descriptions that could also be directed at itself. Edmund Husserl's intentionality made sure that, in all thinking and reflection that is possibly self-referential, a content nevertheless also has to appear indicating an object that is of interest (Husserl, 1970).

Second-order cybernetics generalized the concept of self-reference beyond consciousness and language to include all kinds of autopoietic systems, beginning with cells and organisms through social systems and society (Luhmann, 1990). Yet doubts remain. Humberto R. Maturana and Francisco J. Varela restrict everything said to being said by an observer to an observer (Maturana & Varela, 1980). Heinz von Foerster rather ironically calls for demons, an internal and an external one, to enable a system to self-organize, feeding on the noise of the environment (von Foerster, 2003d; see Dotzler, 1996). And Glanville conceives of the self as one of a trio of self, other, and a modelling facility, which ensures that all distinction again refers to a black box whitening itself by recursively reproducing distinctions relying on nothing but recursion (Glanville, 1991).

Heidegger also struggles with the question where a self may have its place in his joint. For him it is man again who owes and owns a self, the only ones called by and considering the event, which admits the time-space of earth and world, and thus the only one thrown into the openness of be-ing and returning into the—contingent—closure of an entity (Heidegger, 2012, Nos. 127 & 197). Man is the observer who distinguishes the form of entity and be-ing as a form re-entering into itself. To be sure, this man is not to be conceived of in biological, psychological or anthropological terms (Heidegger, Nos. 69, 273 et passim). It is, moreover, an inclusive concept comprehending female, male, transgender, and gender-neutral versions. In fact we are not talking about human beings, but about human be-ings. Asking where in its form this human be-ing may find its place, Heidegger hesitates, considering both language symbolizing man as this operation, more than entity, which finds itself among the objects it collects and constructs, on one hand, and works of art withdrawing themselves as singular instants and thereby flashing opening and closure at the same time, on the other (Heidegger, Nos. 276 & 277f).

My second remark regards the question why Heidegger's joint, Heider's medium, Spencer-Brown's form, Glanville's object or, in fact, Peirce's sign, Shannon's information, Bateson's play and the later Luhmann's system are figures of theory that share not only a peculiar contemporaneity but also a certain emphasis on oscillation. They all oppose the objectivity of the present, the exclusivity of the human subject, and the simplicity of causal explanation to inquire instead into the complexity of the absent being present, the object as well as the subject turning into black boxes and descriptions being instructive of selective interaction. And they all share a more or less radical belief in their originality with respect to older mostly European concepts of being, thinking, time and space. What, if any, is their common historical situation, their common concern (*Geschichte* and *Geschick*)? This is a sociological question, as I must admit. I do not believe in philosophical or theoretical progress as such. A paradigm change—if it is one and, to go by the fierce resistance it is experiencing, it must indeed be one—is signalled by explanations no longer working, new phenomena escaping established concepts, and even feelings missing something that could be important (Kuhn, 1962).

So what is it that triggered such diverse endeavours as pragmatics, semiotics, cybernetics, communication theory, game theory and philosophical deconstruction? I surmise it is to be the electronic revolution presenting us again with the phenomenon of instantaneity (McLuhan, 1964), namely connections at the speed of light we had been used to in tribal society when all communication was oral (by the speed of sound supported, to be sure, by visual perception at the speed of light) yet which had almost been forgotten about when first writing and then the printing press were invented, which introduced us to expanding time horizons we managed only by sticking to a present in a constant attempt to maintain a position from which to assess these horizons. The electronic revolution does not throw us back into the stone age of tribal society, but it does force us to think anew about relations between present and absent, black and white, overt and covert. The figures of theory we are dealing with nowadays try to account for, order, and number energies triggered by minimal information from always uncertain sources (Wiener, 1961). This is why we take note, why we look for joints, forms, and objects, and experimentally place all of these and other elements in networks of dependency and independency (or identity and control; White, 1992), which we barely start to investigate without being referred back to older concepts claiming the authority to tell us what they are.

Acknowledgment

English language editing by Rhodes Barrett.

References

Arendt, H. (1958). *The human condition*. Chicago: Chicago University Press.

Ashby, W. R. (1958). Requisite variety and its implications for the control of complex systems. *Cybernetica, 1*(2), 83-99.

Baecker, D. (1989). Ranulph Glanville und der Thermostat: Zum Verständnis von Kybernetik und Konfusion. *Merkur 43*(484), 513–524.

Baecker, D. (1994). Die Kybernetik unter den Menschen. In P. Fuchs & A. Göbel (Eds.), *Der Mensch—das Medium der Gesellschaft?* (pp. 57-71). Frankfurt: Suhrkamp.

Baecker, D. (2013a). Systemic theories of communication. In P. Cobley & P. J. Schulz (Eds.), *Handbooks of communication science: Vol. 1. Theories and models of communication* (pp. 85-100). Berlin: de Gruyter Mouton.

Baecker, D. (2013b). *Beobachter unter sich: Eine Kulturtheorie*. Berlin: Suhrkamp.

Baecker, D. (in press): Kalkül des Seins. In J. Weiß & G. Tasheva (Eds.), *Existenzialanalystik und Soziologie*. Tübingen, Germany: Mohr Siebeck.

Bateson, G. (2000a). Cybernetic Explanation. In *Steps to an ecology of mind* (pp. 405-416). Chicago: Chicago University Press. (reprint)

Bateson, G. (2000b). A theory of play and fantasy. In *Steps to an ecology of mind* (pp. 177-193). Chicago: Chicago University Press. (reprint)

Dotzler, B. (1996). Demons–magic–cybernetics: On the introduction to natural magic as told by Heinz von Foerster. *Systems Research, 13*(3), 245-250.

Fichte, J. G. (1970). *Grundlage der gesamten Wissenschaftslehre als Handschrift für seine Zuhörer*. Hamburg: Meiner. (Originally published in 1794)

Fichte, J. G. (1986). *Die Wissenschaftslehre: Zweiter Vortrag im Jahre* (R. Lauth & J. Widmann, Eds.). Hamburg: Meiner. (Originally published in 1804)

Freud, S. (1996). *The interpretation of dreams* (A. A. Brill, Trans.). New York: Gramercy Books.

Glanville, R. (1979). The form of cybernetics: Whitening the black box. In B. R. Gaines (Ed.), *General systems research: A science, a methodology, a technology* (pp. 35–42). Louisville, KY: Society for General Systems Research.

Glanville, R. (1980). The architecture of the computable. *Design Studies, 1*(4), 217-225.

Glanville, R. (1982). Inside every white box there are two black boxes trying to get out. *Behavioral Science, 27*(1), 1–11.

Glanville, R. (1987). The question of cybernetics. *Cybernetics and Systems, 18*(2), 99–112.

Glanville, R.(1988). Objekte (D. Baecker, Trans.). Berlin: Merve.

Glanville, R. (1991). The self and the other: The purpose of distinction. In R. Trappl (Ed.), *Cybernetics and systems '90: Proceedings* (pp. 349-356). Singapore: World Scientific.

Glanville, R. (1996). Communication without coding: Cybernetics, meaning and language (How language, becoming a system, betrays itself). *Modern Language Notes, 111*(3), 441-462.

Glanville, R. (1999). Researching design and designing research. *Design Issues, 15*(2), 80-91.

Glanville, R. (2009-2015). The black box (3 vols.). Vienna: edition echoraum

Glanville, R., & Varela, F. J. (1981). Your Inside is Out and Your Outside is in., In G. E. Lasker (Ed.), *International congress on applied systems research and cybernetics* (vol. 6; pp. 638–641). New York: Pergamon Press.

Gunn, J. A. W. (1968). "Interest Will Not Lie": A Seventeenth-Century Political Maxim. Journal of the History of Ideas, 29(4), 551-564.

Heidegger, M. (1977). The end of philosophy and the task of thinking. In D. F. Krell (Ed.), *Basic writings from* Being and Time *(1927) to* The Task of Thinking *(1964)* (pp. 373–392). San Francisco: Harper,

Heidegger, M. (1985). *Schelling's treatise on the essence of human freedom* (J Stambaugh, Trans.) Athens, OH: Ohio University Press.

Heidegger, M. (1995). *The fundamental concepts of metaphysics: World, finitude, solitude* (W. McNeill & N. Walker, Trans.; GA29/30). Bloomington, IN: Indiana University Press.

Heidegger, M. (1998). *Parmenides* (A. Schuwer & R. Rojcewicz, Trans.). Bloomington, IN: Indiana University Press.

Heidegger, M. (2012). *Contributions to philosophy (of the event)* (R. Rojcewicz & D. Vallega-Neu, Trans.). Bloomington, IN: Indiana University Press.

Heider, F. (1959). Thing and medium. In *On perception, event structure, and psychological environment. Selected Papers* (pp. 1–34). *Psychological Issues, 1*(3), New York, NY: International University Press.

Husserl, E. (1970). *The crisis of European sciences and transcencental phenomenology: An introduction to phenomenological philosophy* (Introduction by D. Carr, Trans.). Evanston. IL: Northwestern University Press.

Kauffman, L. H. (1987). Self-reference and recursive forms. *Journal of Social and Biological Structures: Studies in Human Sociobiology 10*(1), 53-72.

Kuhn, T. S. (1962). *The structure of scientific revolutions*. Chicago: Chicago University Press.

Locke, J. (1959). *An essay concerning human understanding* (2 vols.; A. C. Fraser, Ed.) New York: Dover.

Luhmann, N. (1990). *Essays on self-reference*. New York: Columbia University Press.

Luhmann, N. (1997). The control of intransparency. *Systems Research and Behavioral Science, 14*(6), 359-371.

Luhmann, N. (2002a). The modern sciences and phenomenology. In W. Rasch (Ed.), *Theories of distinction: Redescribing the descriptions of modernity* (pp. 33-60). Stanford, CA: Stanford University Press.

Luhmann, N. (2002b). The cognitive program of constructivism and the reality that remains unknown. In W. Rasch (Ed.), *Theories of distinction: Redescribing the descriptions of modernity* (pp. 128–152). Stanford, CA.: Stanford University Press.

Luhmann, N. (2012/13). *Theory of society* (2 vols.; R. Barrett, Trans.) Stanford, CA: Stanford University Press.

Marx, K., & Engels, F. (1976). *The German ideology: Including "Theses on Feuerbach" and "Introduction to the critique of political economy"* (C. Dutt, W. Lough & C. P. Magill, Trans.). Amherst, NY: Prometheus Books.

Maturana, H. R., & Varela, F. J. (1980). *Autopoiesis and cognition: The realization of the living*. Dordrecht: Reidel.

McLuhan, M. (1964). *Understanding media: The extensions of man*. New York McGraw-Hill.

Mejía, E., & Schüßler, I. (Eds.) (2009). *Heidegger's "Beiträge zur Philosophie."* International Colloquium, May 20-22, 2004, University of Lausanne. Frankfurt: Klostermann.

McCulloch, W. S. (2004). The Beginning of Cybernetics, In C. Pias (Ed.), *Cybernetics: The Macy-Conferences 1946–1953: Vol. 2. Essays and Documents* (pp. 345–360). Zürich: diaphanes.

Nietzsche, F. (2003). *Writings from the late notebooks* (R. Bittner, Ed. K. Sturge, Trans.). Cambridge: Cambridge University Press.

Parsons, T. (1977). Social structure and the symbolic media of interchange. I*n Social systems and the evolution of action theory* (pp. 204-228). New York: Free Press.

Plato. (1993). *Sophist* (Introduction and notes by N. P. White, Trans.). Indianapolis, IN: Hackett.

Pöggeler, O. (1997). *The paths of Heidegger's life and thought* (J. Bailiff, Trans.). Amherst, NY: Prometheus Press.

Polt, R. F. H. (2006). *The emergency of being: On Heidegger's "Contributions to Philosophy."* Ithaca, NY: Cornell University Press.

Shannon, C. E., & Weaver, W. (1963). *The mathematical theory of communication*. Urbana, IL: Illinois University Press. (Reprint)

Spencer-Brown, G. (2008). *Laws of form* (5th international ed.). Leipzig, Germany: Bohmeier.

Thomä, D. (1990). *Die Zeit des Selbst und die Zeit danach: Zur Kritik der Textgeschichte Martin Heideggers 1910-1976*. Frankfurt: Suhrkamp.

von Foerster, H. (1971). Computing in the semantic domain. *Annals of the New York Academy of Sciences, 184*, 239–241.

von Foerster, H. (2003a). Thought and Notes on Cognition, In *Understanding understanding: Essays on cybernetics and cognition* (pp. 169-189). New York: Springer.

von Foerster, H. (2003b). On constructing a reality. In *Understanding understanding: Essays on cybernetics and cognition* (pp. 211-227). New York: Springer.

von Foerster, H. (2003c). Notes on an epistemology for living things. In *Understanding understanding: Essays on cybernetics and cognition* (pp. 247-259). New York: Springer.

von Foerster, H. (2003d). *On self-organizing systems and their environment.* In *Understanding understanding: Essays on cybernetics and cognition* (pp. 1-20). New York: Springer.

von Glasersfeld, E. (1995). *Radical constructivism: A way of knowing and learning.* London: Falmer.

von Uexküll, J. (1926). *Theoretical biology.* New York: Harcourt, Brace & Co.

Wertheimer, M. (1925). *Drei Abhandlungen zur Gestalttheorie.* Erlangen, Germany: Verlag der Philosophischen Akademie.

White, H. C. (1992). Identity and control: A structural theory of action. Princeton, NJ: Princeton University Press.

Wiener, N. (1936). The rôle of the observer. *Philosophy of Science, 3*(3), 307-319.

Wiener, N. (1961). *Cybernetics, or control and communication in the animal and the machine* (2nd ed.). Cambridge, MA: The MIT Press.

Wittgenstein, L. (2001). *Tractatus Logico Philosophicus.* London: Routledge.

Bunnell, P. (2010). *Tidal Abode.* Un-retouched photograph.

Cybernetics and Human Knowing. Vol. 22 (2015), nos. 2-3, pp. 59-71

Putting Flesh on the Bones:
Ranulph Glanville's Contributions to Conversation Theory

Bernard Scott[1]

Gordon Pask's conversation theory stands as a major contribution to cybernetics, education and epistemology. Ranulph Glanville was Pask's student. He made himself familiar with conversation theory and promoted it in his lecture and writings throughout his career. In doing so, he made significant contributions. I summarise some of these here. To set the scene, I briefly outline some of the main concepts of conversation theory. I also outline Glanville's theory of Objects and discuss its relationship to conversation theory.

Keywords: conversation, conversation theory, theory of Objects, cybernetics, second-order cybernetics, design

Introduction

The UK cybernetician, Gordon Pask (1928-1996), was responsible for a large body of empirical and theoretical work spread over several decades. His core ideas were summed up in his theory of conversations (conversation theory [CT]) and his interaction of actors (IA) theory. IA theory refers to later elaborations of CT which I will not deal with separately here. My particular concern is with Pask's classic statements of CT as set out in Pask, Scott and Kallikourdis (1975) and Pask (1975b, 1976). As emphasised in his final paper (Pask, 1996), CT is concerned with the interactions of self-organising systems and the particular case where the systems in question are one or more human beings and an adaptive teaching machine.[2] Pask sets out in detail the logic of these interactions, which he refers to as conversations, as a set of descriptions of what takes place when conversation can be considered to be fruitful and effective and as a set of prescriptions for how one can ensure that conversations are indeed fruitful and effective. Ranulph Glanville, to whom this paper is a tribute, was one of Pask's students. Indeed, it is fair to say that he was one of Pask's foremost students. He was the first student to gain a PhD in Cybernetics from Brunel University (in 1975) and did so under Pask's tutelage. Glanville was studying with Pask at a time when Pask's CT was coming to fruition as a coherent and technically complex set of concepts that, amongst other topics, addressed the dynamics of learning, individual differences in learning style and the structure of what Pask referred to as "conversational domains," the bodies of knowledge that learners come to know and that teachers attempt to teach. In addition to his own original work that he carried out for his PhD studies (of which I will say more below), Glanville rapidly embraced and

---

1. Email: Bernces1@gmail.com
2. For an account of Pask's extensive work on adaptive teaching, see Pask (1975a).

familiarised himself with Pask's CT. He became a champion of CT. He considered it to be a major contribution to cybernetics and to the broad domains of education and epistemology and promoted it as such in his own lectures and writings. He helped make CT available and accessible to many who found Pask's own writings and expositions to be conceptually and technically daunting. Pask's CT continued to play a central role in Glanville's thinking throughout his career, as he developed his own original theories of design and design education and produced many original contributions to the cybernetics literature.

In this paper I summarise Glanville's contributions to CT and their significance as I see them. I have chosen the metaphor in the title, "putting flesh on the bones," quite deliberately. At the core of CT is a logical structure, which Pask calls "the skeleton of a conversation." Thus, as per my title, I wish to show how Glanville adds the flesh to the bones of Pask's conversational skeleton and brings it to life as a rich portrayal of what happens and what may happen when human beings converse.

The paper is structured as follows: First I give an account of the conversational skeleton: the logical structure of CT. Next I give a brief account of Glanville's theory of Objects, the topic of Glanville's PhD thesis. My intention here is to show how in his own theorising Glanville is alive to the kinds of distinction that need to be made in the construction of a theory of conversations. I then address the main topic of my paper: a description of how, in his commentaries on CT, Glanville made many valuable and enriching contributions to CT. I also briefly make mention of Glanville's long term interest in the processes of design and his use and elaboration of CT in that context.

In the final part of the paper, I attempt to set these contributions of Glanville's within the larger context of second-order cybernetics. In doing so, I will make brief mention of other ways Glanville has contributed to cybernetics.

Conversation Theory

CT grew from Pask's interests in adaptive teaching systems.[3] He argued that the interaction between a learner and a teaching machine has the form of a conversation. Pask developed CT from first principles in order to formalise his ideas.[4] By this, I mean he begins by making clear what the distinctions are that an observer of a conversation must make in order to describe and characterise that conversation. These are:

(i) A distinction between participants, capable of making self and other reference
(ii) A distinction between the cognitive processes (the processes of knowing) that constitute the participants as psychological individuals (Pask refers to these as P-

---

3. A non-technical account of the development of conversation theory can be found in Scott (1993).
4. CT can be considered as a general theory of learning and teaching with particular relevance for e-learning applications (Scott, 2001). In his comprehensive and detailed survey of web-based learning, Nigel Ford (2008) uses CT as a unifying conceptual framework and reports extensively on the empirical studies carried out by Pask and Scott.

Individuals) and the medium or fabric that embodies those processes (Pask refers to these embodiments as M (mechanical)-individuals). For humans, the relevant M individuals are brain/body systems and their augmentations.

(iii) A distinction between the participants and the shared world in which they interact behaviourally (Pask refers to this shared world as a modelling facility).

(iv) A distinction between the cognitive processes that bring about, maintain, recognise or recall relations (events, processes) in the modelling facility using the resources of that facility and the cognitive processes that bring about, maintain, recognise or recall cognitive processes using the resources of the brain/body system and its augmentations. Pask's general term for a cognitive process is a *concept*. In early writings on CT, Pask uses the term *memory* for concepts that act on or construct other concepts. He characterised a unitary P-Individual as a self-reproducing system of memories and concepts.

(v) A distinction between the behavioural interactions between participants as they use the shared modelling facility to construct and co-construct particular models and the symbolic (verbal, linguistic) interactions by which the participants attempt to modify each other's behaviours and cognitive states. Pask refers to these latter interactions as *provocations*. Provocations take place via a suitable interface, one that affords the making of speech acts, writing or graphical inscriptions according to agreed semantic and syntactic rules. The pragmatics are provided by the roles that the participants have agreed to play as part of their contract with the experimenter. In a typical experiment run by Pask and colleagues, one participant would act as a learner, the other would act as a teacher. Frequently, the teacher role was filled by a teaching machine, programmed to ensure that the learner learned effectively. The important thing to appreciate is that the interface needs to be such that the external observer can observe the provocations that are taking place between the participants. In Pask's experimental setup for observing conversations (Course Assembly System and Tutorial Environment [CASTE]), the participants were required to use the interface for their communications. The experimenter could then observe what was taking place.[5]

These five distinctions allow Pask to construct his "skeleton of a conversation," shown in Figure 1. Figure 1 shows the situation in which one participant is a teacher conversing with a second participant the learner and in which the learner has agreed to attempt to come to an understanding of the thesis that the teacher professes. Pask referred to a tutorial conversation of this kind as a *strict conversation*. In a strict conversation, for the benefit of the experimental observer, the domain of the conversation is fixed: the teacher has a thesis that the learner is contracted to learn.

The formalities of a strict conversation are a long way away from what happens in natural conversations. In less strict conversations, the domain of a conversation may shift as participants take turns to lead the conversation from topic to topic. Indeed, the

---

5. For details of CASTE, see Pask and Scott (1973), Pask (1975) or Scott (2001).

conversation itself may become a topic of conversation. Participants may go on to negotiate or re-negotiate their roles. Pask was well aware of these and other possible elaborations and in due course added to CT to take them into account (Pask, 1976).

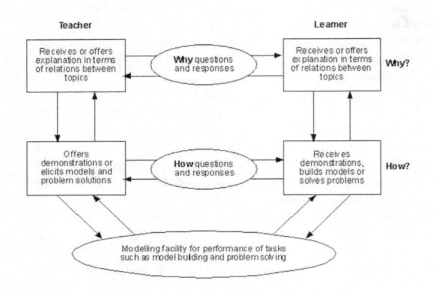

Figure 1. The Skeleton of a Conversation (after Pask).

Figure 1 shows a snapshot view of two participants in conversation about a topic. The horizontal connections represent the provocative exchanges. Pask argues that all such exchanges have, as a minimum, two logical levels. In the figure these are shown as the two levels: How and Why. The How level is concerned with descriptions of how to do a topic: how to recognise it, construct it, maintain it and so on; the Why level is concerned with explaining or justifying what a topic means in terms of other topics. These exchanges are provocative in that they serve to provoke participants to construct understandings of each other's conceptions and (possibly) misconceptions of topics and the relations between them. This is the essential aspect that makes conversation theory constructivist and dialogical in approach and clearly distinguishes it from other approaches that see teaching as the transmission of knowledge from teacher to learner.

The vertical connections represent causal connections with feedback, a hierarchy of cognitive processes that control or produce other cognitive processes. At the lowest level in the control hierarchy there is a canonical world, a *universe of discourse* or *modelling facility* where the teacher (or computer-based surrogate, as incorporated in CASTE) may instantiate or exemplify the topic by providing non-verbal demonstrations. Typically, such demonstrations are accompanied by expository narrative about How and Why, the provocative interactions of questions and answers referred to above. Note that the form of what constitutes a canonical world for construction and demonstration may itself be a topic for negotiation and agreement.

Consider, for example, a set of well-defined topics in chemistry. A teacher may:

- model or demonstrate certain processes or events;
- offer explanations of why certain processes take place;
- request that a learner teaches back his or her conceptions of why certain things happen;
- offer verbal accounts of how to bring about certain events;
- ask a learner to provide such an account; and
- ask a learner to carry out experiments or other practical procedures pertaining to particular events or processes.

A learner may:

- request explanations of why;
- request accounts of how;
- request demonstrations;
- offer explanations of why for commentary;
- offer explanations of how for commentary; and
- carry out experiments and practical activities.

In turn, the learner uses the modelling facility to solve problems and carry out tasks set. He or she may also provide narrative commentary about How and Why. In a computer-based environment these may be elicited using computer aided assessment tools with a variety of different question styles. The distinction between How and Why allows for a formal definition of what it means to understand a topic. In CT, understanding a topic means that the learner can *teachback* the topic by providing both non-verbal demonstrations and verbal explanations of How and Why.

Pask notes that conversations may have many levels coordination above the Why level: levels at which conceptual justifications are themselves justified and where there is commentary about commentary. Harri-Augstein and Thomas (1991) make this notion central in their work on self-organised learning, where the emphasis is on helping students learn to learn. In brief, they propose that a full learning conversation has three main components:

(1) conversation about the how and why of a topic, as in the basic Pask skeleton of a conversation model;
(2) conversation about the how of learning (for example, discussing study skills and reflecting on experiences as a learner); and
(3) conversation about purposes, the why of learning, where the emphasis is on encouraging personal autonomy and accepting responsibility for one's own

learning.

As we shall see below, Granville helps bring the skeleton of a conversation to life (to put flesh on its bones) and allows us to see how CT captures the essence of what is happening when humans converse. Glanville brings this process of conversing one with another, of learning from and with each other, to life and thus emphasises the creativity and joyfulness that can, indeed should under ideal circumstances, accompany human conversation.

It would be remiss not to mention another facet of Pask's CT that is central to his concerns. This is his detailed account of *conceptualisation*[6] the processes whereby new concepts come into being amidst the flux of the ongoing evolutionary dynamics of the self-reproducing system of memories and concepts that constitute a P-Individual.[7] Pask's insight, inspired by the work of Lev Vygotsky (1962)[8] is to realise that these processes are conversational in form. There is an inner dialogue in which a learner constructs, contemplates and contrasts different perspectives of a topic: her own, her teacher's, her peers or other authoritative sources. It is this inner dialogue that permits the effective learner to act as her own teacher. With this insight, Pask offers a unifying perspective on individual and social psychologies: a p-individual is conversational in form; a conversation has the form of a p-individual. Both are examples of *psychosocial unities* (Scott & Bansal, 2014). I refer again to the concept of an inner dialogue, below, in my discussion of Glanville's contributions to CT.

Glanville's Theory of Objects[9]

Glanville's PhD thesis (Glanville, 1975) addresses the following question: How is it that observers, all of whom observe and know differently from each other, come to believe, and behave as if, they are observing thing the same thing? In developing his answer to this question, he recognised that any entity, to have the stability that constitutes it as an entity, has to remember itself, to reconstruct itself as itself. Memory is the process that has itself as a product. With one stroke Glanville places circularity and self-reference at the very heart of an abstract cybernetic algebra. Expressed in later terminology, Glanville developed a second-order cybernetics (of observing systems) and at the same time established that all first order forms (observed systems) intrinsically carry with them second-order considerations (all observed systems are observing systems; all systems are, mutually, self-observing).[10] In his reading, Glanville had discovered that the word *object* had once upon a time had the connotation that we now give to the word *subject*. As a quirky twist on this historical

---

6. In some earlier publications, Pask uses the term *mentation* to refer to these processes. See, e.g., Pask (1968).
7. For a detailed discussion of these dynamics, modelled as a computer simulation, see Scott and Bansal (2014).
8. Pask acknowledges his long standing indebtedness to Vygotsky in his final paper, Pask (1996).
9. This brief account of Glanville's theory of Objects is adapted from Scott (2005).
10. The distinction between first-order and second-order cybernetics is due to Heinz von Foerster (Foerster et al, 1974).

change of meaning, Glanville refers to his self-observing entities that are both subject and object for themselves as *Objects*.

Glanville gives an account of his own contribution thus:

> My work might be thought of as a generalization of the work of the others. My major initial concern was to develop a set of concepts that might explain how, while we all observe and know differently, we behave as if we were observing the same thing. What structure might support this? (Glanville, 2012, p. 192)

Notice that Glanville is concerned to acknowledge the contributions of peers and mentors. He later lists the key players as Gordon Pask, Heinz von Foerster, Humberto Maturana and Francisco Varela. He continues:

> My contribution was a structure developed to accommodate observation and difference. This was achieved by arguing mutualism, here glossed as "the reciprocal arrangement by which what may be of one may be of the other". When drawing a distinction that which can be assumed for one side must in principle at least be possible for the other. This I have called the "Principle of Mutual Reciprocity."(Glanville, 2012, p. 192)

Here, I believe, is the key idea that does indeed generalise over the work of others. The reader should appreciate that the Principle of Mutual Reciprocity is pre-ontological in that, as yet, there is no commitment to a particular shared reality. If an *I* distinguishes a *Thou*, what is allowed of the one is allowed of the other. If an *I* distinguishes a *me* and a *he, she* or *it*, what is allowed of the one is allowed of the other. Glanville goes on:

> In a universe of discourse determined by individuality and difference in observation, observing entities are taken to observe themselves: they are self-referential. Thus they attain identity and autonomy. (Observation should not be confused with seeing: observation as used here is a formal quality.) Therefore, observed entities must be assumed to have the possibility that they observe themselves. (Glanville, 2012, p. 192)

Here, we are still pre-ontological but we are obliged by the Principle of Mutual Reciprocity to admit of the observed that which we have admitted of the observer: the property of self-observation, giving identity and autonomy. Glanville refers to this class of self-observing entities as Objects, as in the following extract:

> It is considered inconceivable that such entities (called "Objects") are simultaneously both self-observing and self-observed. They are therefore taken to switch roles. This generates time (making time a central and integral concept in second order Cybernetics), allows observation by another Object, and sets up observational time as a way of relating observations of other Objects, giving a relational logic. Objects are seen as oscillating between the two roles, and this oscillation allows the continuity of the observation of self; and the observation of others in time, giving rise to relationships. Objects generate process, just as they are generated by process: another cybernetic circularity. Since observation can thus take place, it is assumed other activities can also occur. (Glanville, 2012, p. 192-193)

Glanville has here posited another key pre-ontological principle, which could be named as "the Principle of the Exclusion of Observing and Being Observed." This is first established for the *I* and the associated *me* of one Object and then extended to the case of Objects observing each other. The exclusion principle brings forth the concept of time as the difference between observing and being observed and also brings forth the concept of process, something happening. An Object is seen to be an entity that as a process constitutes itself as itself. It is the memory of itself.

This is the beginning of Glanville's account. Elaborations go on to set out the full structure for the formation of higher orders of self-reflection and for the formation of temporally synchronised social groups and coalitions.[11] It should be emphasised once more that this structure is pre-ontological. As yet there are no laws of physics, there are no forms designated living or non-living, there is no beginning or ending. Glanville (2002) sums up:

> To use a metaphor: my work is the creation of games fields: others create the games to play in these fields and still others play them. Finally, some are spectators. The point of an account that admits others is not that it is right, but that it is general (and generous). Cybernetics is often considered a meta-field. The Cybernetics of Cybernetics is, thus, a meta-meta-field. My work is, therefore, a meta-meta-meta-field. (Glanville, 2012, p. 193)

What then has Glanville achieved? His structure allows us to play, to observe and be observed, to construct maps and models and ontologies to our hearts' contents. It also reminds us that this form, the form of playing, is indeed pre-ontological and it behoves us to act accordingly with respect to one another. In his *Philosophical Investigations*, Ludwig Wittgenstein (1953) characterise philosophy as a ground clearing exercise. By an exercise in abstract cybernetic algebra, Glanville has cleared the primordial ground. It is from this ground that all theories, first and second order, are built, with Pask's CT as an example of a second-order theory of theory building.

Glanville's Contributions to Conversation Theory

Glanville wrote about conversation theory and the life and work of Gordon Pask on many occasions. Here, I draw in particular upon Glanville (1993, 2001, 2002). What Glanville recognised was that hidden beneath the complex technical apparatus of Pask's theory there is "a great humanist undertaking" (Glanville, 2001, p. 661) addressing fundamental issues about what it means to be human and to interact (converse) one with another. As a tribute to Pask, Glanville sets out a scheme that elaborates "the qualities necessary so that a conversation may function" (Glanville, 2001, p. 661).

> To create such a scheme is, I believe, the greatest tribute I can pay to his understanding, his insight, his work and his imagining. It also throws light on us as people: on our behaviour and our

---

11. These elaborations were published in a series of papers, later collected together, translated into German (by Dirk Baecker) and published in book form with the title *Objekte* (Glanville, 1988).

inspiration, and on the ethics which were always behind Gordon's work but which (as with other "philosophical" issues) he seemed so reluctant to put into explicit written form. (Glanville, 2001, p. 661)

Here, I outline Glanville's scheme with some additional commentaries of my own.

Glanville refers to two sets of requirements for a conversation to function, which he refers to as *operational requirements* and *inspirational requirements*. By the former he means those aspects of the interaction process which must be present for the interaction to be considered to be a conversation, rather than a more arbitrary or limited encounter between participants such as a simple exchange of greetings, goods or physical contacts. One could perhaps also refer to them as functional or processual requirements. Inspirational requirements concern the attitudes and motivations that it is necessary for participants to bring to the conversation for it to flourish as a mutually creative and uplifting encounter. Both sorts of requirements take the form of tacit, sometimes explicit, reciprocal expectations and are akin to the *cooperative principle* of Paul Grice that states, "Make your contribution such as it is required, at the stage at which it occurs, by the accepted purpose or direction of the talk exchange in which you are engaged" (Grice, 1975, p. 45).[12] A key difference is that Glanville's requirements are concerned with the whole of the encounter that is the conversation, whereas Grice is concerned at a more micro-level with good (effective) practice with respect to individual speech acts and responses and the implications (conversational implicatures) that can be drawn from them on the assumption that the cooperation principle is being adhered to.

Glanville's operational requirements:[13]

(1) A willingness to take part in a conversation about some topic. At least two participants are needed.

(2) The topic around which the conversation takes place. The topic is negotiable and may change. There is an ever present background topic, the reflexive topic, "What shall we talk about?"

(3) The existence of different understandings of the topic in all participants. Without these differences, there would be no need for conversation.

(4) Acts that are intended to present the form of these understandings so that the other participants can construct their own understandings of these understandings together with acts that are intended to request the presentation of understandings (questions).

(5) An ability to compare understandings: my understanding with my

---

12. Grice goes on to state a set of maxims to be followed if the cooperative principle is accepted: Make your contribution as informative as is required (for the current purposes of the exchange). Do not make your contribution more informative than is required Do not say what you believe to be false. Do not say that for which you lack adequate evidence. Avoid obscurity of expression. Avoid ambiguity. Be brief (avoid unnecessary prolixity). Be orderly.

13. For the sake of brevity, I have paraphrased Granville's statements and omitted some of his elaborations. I have done this for both sets of requirements.

understanding of your understanding of my understanding and, vice versa,
your understanding with your understanding of my understanding of your
understanding. (In Pask's terms, this is part of the inner dialogue that is
informs conceptualistion.)

(6) A logical structure of three co-located and contemporaneous levels: the level
of the conversation, the subordinate level of the topic being addressed, and the
metalevel of error correction and topic modification (critique and evaluation
of the conversation's progress).

(7) An ability to monitor what is going on and to correct for incompatibilities
between understandings by switching levels, that is:

- switching to a meta-conversation in which misunderstandings
  (temporarily) become the topic of the conversation
- switching back to the topic of the conversation itself. (Glanville notes that
  these switchings can occur recursively: misunderstandings may be
  misunderstood and require further error correction.)

(8) A way of initiating and terminating the conversation. (Glanville notes that in
Pask's CT, the occurrence of an understanding punctuates the conversation
into discrete, possibly concurrent, episodes. In real life conversations,
participants may terminate a conversation when mutual understanding is
acknowledged, when there is an agreement to disagree or as a matter of whim.
Of course, the conversation may be taken up again on a future occasion.
Glanville further notes that in real life conversations, confirmation of mutual
understandings is not expected for each and every communicative act. This is
especially the case when participants believe that they already share many
understandings. However, there is a price to pay for these shortcuts: the
inadvertent pathologies of communication that occur when everyone thinks
everyone else knows what is going on, when, in fact, they do not.[14]

Glanville's inspirational requirements

(1) Recognition that the other has a different understanding.
(2) Respect for this difference and the owner of the difference. Respect allows the
participants to form their own individuality. Respect allows that I am not you.
(3) Willingness to listen and hear the other.
(4) Willingness to construct my own understanding of what the other presents to
me as her understanding.
(5) Willingness not to try to force my view on the other, that is, not to exploit
power relationships due to differences in social position.
(6) An open mind, that is, being prepared to give space to the other and to
negotiate.
(7) To regard surprises in the conversation not as threats but as being beneficial as

---

14. For a discussion of these pathologies and the forms they take in social organisations, see Scott (1997).

opportunities to learn.

(8) Willingness to change, develop, improve, that is, to learn.

(9) To recognise that what arises in conversation is not the property of a particular individual participant but rather is jointly owned. This is to recognise that the conversation has a life of its own. (In Pask's terms, the conversation itself is a p-individual),

(10) A willingness to go with conversation, to expect and allow for the unexpected.

Glanville argues that underlying these inspirational requirements are certain qualities that are associated with being a good and decent human being. These qualities are generosity, respect, honesty and a sense of drama; openness, imagination, acting on opportunities and wit. He ascribes these qualities to his mentor, Gordon Pask, and goes on to say of his encounters with him:

> This is magic. Magic not as trickery or deceit, but magic in the unravelling enjoyment of mysteries and the growing and maintaining of wonder a deep understanding of the miracle of our existing in our differing worlds and of their coming together in conversation through their beginnings and ends. Of the poetic nature of our existence and of the unity of the void, the nothingness in and through which we dwell. And the love that is necessary that we can converse and interact with those others with whom we dwell, fairly and doing justice to them and to ourselves. (Glanville, 2001, p. 667)

Glanville on the Processes of Design

As many readers will know, the processes of design were a central and long term interest for Glanville as a theorist, practitioner and teacher.[15] He took his inspiration from Pask's CT, in particular, Pask's characterisation of conceptualisation as an inner conversation. Glanville elaborated upon this, noting how creative designers typically externalise aspects of their conceptualisation in the form of drawings or other artefacts and then go on to engage in conversations with themselves about those artefacts, in a sense, allowing the artefacts to be a source of new ideas. Glanville emphasised that these processes cannot be proceduralised as a set of steps to be followed. They require qualities similar to those required when engaging in effective external conversations: generosity and respect towards alternative perspectives, honesty (in this case with oneself), openness, imagination, acting on opportunities and wit. Glanville generalised the idea of design to encompass all attempts by humans to construct aesthetically pleasing, rewarding and joyful ways of life. Apropos of this broader notion of design, Granville proposed that in second order cybernetics one could usefully substitute the term *composer* for the much used and abused term *observer* in order to emphasise the active and creative aspects of being human.[16]

---

15. See, as an example of his thinking on this topic, Glanville (2007).

Concluding Comments

In setting out the technicalities of Pask's skeleton of a conversation and the abstract metaphysics of Glanville's theory of objects, I have tried to show the cybernetic nature of their thought, in the spirit of Ross Ashby's statement that cybernetics has its own foundations and "takes as its domain of subject matter 'all possible machines'" Ashby (1956, p. 2). I have tried to show the original contributions that Glanville has made to second-order cybernetics and, in particular, to Pask's conversation theory. Taking inspiration from Glanville, I have stressed the deep humanism that I believe pervades all cybernetic thought.[17] This humanism is to be found not only in Pask and Glanville but also in the lives and works of other great cyberneticians, as examples: Warren McCulloch, Ross Ashby, Gregory Bateson, Margaret Mead, Heinz von Foerster, Gotthard Guenther, Stafford Beer and Humberto Maturana. As this paper is a tribute to the life and work of Ranulph Glanville, I wish to emphasise how evident it is to all who knew him that Ranulph lived his cybernetics wholeheartedly and with great commitment. As a freelance academic, he had the courage to promote cybernetics long after it had become unfashionable and helped revive its fortunes.[18] He was extraordinarily generous towards his peers and his students. He gave generously of himself to the furthering of the field of cybernetics, not least in what he did for the American Society for Cybernetics and what he did in creating and fostering the Cybernetics Coalition. To be with him was to be inspired, to experience wonder and to have fun. Ranulph was always ready to disarm us with his honesty, his endless supply of puns and his love.

References

Ashby, W. R. (1956). *Introduction to cybernetics*. New York: Wiley.
Foerster, H. von et al. (Eds.). (1974). *Cybernetics of cybernetics*. BCL Report 73.38. Urbana, IL: Biological Computer Laboratory, Dept. of Electrical Engineering, University of Illinois.
Ford, N. (2008). *Web-based learning through educational informatics*. Hershey, PA: Information Science Publishing.
Glanville, R. (1975). *The object of objects, the point of points—or, Something about things*, Ph.D. thesis, Brunel University, London.
Glanville, R. (1993). Pask: A slight primer. *Systems Research, 10*(3), 213-318.
Glanville, R. (1988). *Objekte* (D. Baecker, Trans.). Berlin: Merve Verlag.
Glanville, R. (2001). And he was magic. *Kybernetes, 30*(5/6), 652-672.
Glanville, R. (2007). Try again. Fail again. Fail better. The cybernetics in design and the design in cybernetics. In R. Glanville (Ed.), *Cybernetics and Design* [Special issue]. *Kybernetes, 36*(9/10), 1173-1206.
Glanville, R. (2012).*The black b∞x: Vol 1. Cybernetic Circles* (Chapter 1.11; pp. 175-207). Vienna: Echoraum-WISDOM.
Glanville, R. (2014). Cybernetics: thinking through the technology. In D. Arnold (Ed.), *Traditions of systems theory: Major figures and contemporary developments* (pp. 45-77). New York: Routledge.

---

16. For recent lecture by Glanville that covers these topics in an accessible, non-technical way, see https://www.youtube.com/watch?v=Z8g7GA6DEU8 (published 2914, accessed 24/01/15). This is a recording of Glanville's Professorial inaugural lecture, an event in the Bartlett School of Architecture's International Lecture Series 2009/2010, delivered on the 10th of March 2010, with introductory remarks by Marcos Cruz and Professor Stephen Gage.
17. So much so that I have described cybernetics as "the art and science of fostering goodwill" (Scott, 1993, p.167).
18. See Glanville (2014) for a masterly example of these efforts.

Grice, P. (1975). Logic and conversation. In P. Cole & J. Morgan (Eds.), *Syntax and semantics: Vol. 3. Speech acts* (pp. 41-58). New York: Academic Press.

Harri-Augstein, S. & Thomas, L. F. (1991). *Learning conversations*. London: Routledge.

Pask, G. (1968). A cybernetic model for some types of learning and mentation. In H. C. Oestreicher & D. R. Moore (Eds.), *Cybernetic problems in bionics* (pp. 531-585). New York: Gordon and Breach.

Pask, G. (1975a). *The cybernetics of human learning and performance*, London: Hutchinson.

Pask, G. (1975b). *Conversation, cognition and learning*. Amsterdam: Elsevier.

Pask, G. (1976). *Conversation theory: Applications in education and epistemology*. Amsterdam: Elsevier.

Pask G. (1996). Heinz von Foerster's self-organisation, the progenitor of conversation and interaction theories. *Systems Research, 13*(3), 349–362.

Pask, G., & Scott, B. (1973). CASTE: A system for exhibiting learning strategies and regulating uncertainty. *Int. J. Man-Machine Studies, 5*, 17-52.

Pask, G., Scott, B., & Kallikourdis, D. (1973). A theory of conversations and individuals (exemplified by the learning process on CASTE). *Int. J. Man-Machine Studies, 5*, 443-566.

Scott, B. (1993). Working with Gordon: Developing and applying conversation theory (1968-1978). *Systems Research, 10*(3), 167-182.

Scott, B. (1997). Inadvertent pathologies of communication in human systems. *Kybernetes, 26*(6/7), 824-836.

Scott, B. (2001). Conversation theory: A dialogic, constructivist approach to educational technology. Cybernetics and Human Knowing, *8*(4), 25-46.

Scott, B. (2005). Ranulph Glanville's *Objekte*: An appreciation. In D. Baecker (Ed.), *Schluesselwerke der Systemtheories*. Wiesbaden: VS Verlag für Sozialwissenschaften.

Scott, B., &Bansal, A. (2014). Learning about learning: A cybernetic model of skill acquisition. *Kybernetes, 43*(9/10), 1399-1411.

Vygotsky, L. (1962). *Thought and language*. Cambridge, MA: The MIT Press.

Wittgenstein, L. L. (1953). *Philosophical investigations*. Oxford, UK: Basil Blackwell.

Bunnell, P. (2010). *Silver and Gold*. Un-retouched photograph.

Bunnell, P. (2014). *Conceptual Cave*. Un-retouched photograph.

Cybernetics and Human Knowing. Vol. 22 (2015), nos. 2-3, pp. 73-82

Cybernetics and Design:
Conversations for Action

Hugh Dubberly[1] and Paul Pangaro[2]

Working for decades as both theorist and teacher, Ranulph Glanville came to believe that cybernetics and design are two sides of the same coin.

Working as both practitioners and teachers, the authors present their understanding of Glanville and the relationships between cybernetics and design.

We believe cybernetics offers a foundation for 21st-century design practice. We offer this rationale:

- If design, then systems: Due in part to the rise of computing technology and its role in human communications, the domain of design has expanded from *giving form* to *creating systems* that support human interactions; thus, systems literacy becomes a necessary foundation for design.
- If systems, then cybernetics: Interaction involves goals, feedback, and learning, the science of which is cybernetics.
- If cybernetics, then second-order cybernetics: Framing wicked problems requires explicit values and viewpoints, accompanied by the responsibility to justify them with explicit arguments, thus incorporating subjectivity and the epistemology of second-order cybernetics.
- If second-order cybernetics, then conversation: Design grounded in argumentation requires conversation so that participants may understand, agree, and collaborate on effective action.

Second-order cybernetics frames design as conversation for learning together, and second-order design creates possibilities for others to have conversations, to learn and to act.

Keywords: design, systems, cybernetics, second-order cybernetics, conversation, ethics, language, models, design knowledge, design methods, design rationale

---

A Conversation about Conversations-for-Action

In October of 2014, the authors began a conversation with Ranulph Glanville about the relationships between cybernetics and design. Our all-too-brief conversation with him is the basis for this paper. We should acknowledge that this paper is not a review of Glanville's extensive writings and that we may not fully understand his views. However, we would like to report on the points he made, sometimes quite vehemently, to us—and we would like to comment on the many places where we concur and the few where we do not.

The catalyst for our conversation was Glanville's masterful presentation at the RSD3, Relating Systems Thinking and Design 2014 Symposium in Oslo (Glanville, 2014a). Glanville argued that first-order cybernetics, far from being mere mechanics or calculation, provides a necessary alternative to linear causality: It brings us circular causality, critical to understanding and realizing (making) interactive systems that evolve through recursion, learning, and co-evolution. Second-order cybernetics is

---

1. Dubberly Design Office, 2501 Harrison Street, San Francisco, CA 94110 USA. Email: hugh@dubberly.com
2. College for Creative Studies, 201 E. Kirby St., Detroit, MI 48202 USA. Email: paul@pangaro.com

fundamental to design because it gives us an epistemological framework for designing.[3] Second-order cybernetics moves us from a detached, "objective" pose where we can duck responsibility, and right into the messy middle of things, where only we can be responsible for our actions.

Second-order cybernetics frames design as conversation for learning together. This creates the conditions for better-directed, more deliberate outcomes: hence the second half of our title, "Conversations for Action."

Sadly, Glanville's passing cut short our conversation. We strive to present his views as best we understand them, quoting him when possible. We appreciate his gifts, and we miss him. We invite continued conversation, especially with others who have collaborated with him and who may see his intentions differently. Together let us evolve the field.

The Context for Cybernetics and Design

We construe design in the broad sense as a conversation for action—that is, as cybernetics. Action may either conserve or change, a conversation about what we value. In other words, design is a conversation about what to conserve and what to change. So is a cybernetic system. Both involve a process of observing a situation as having some limitations, reflecting on how and why to improve that situation, and acting to improve it. This follows the circular process of observe→reflect→make that is common to the recursive and accumulative process of learning in service of effective action, as is found in science, medicine, biological systems, manufacturing, and everyday living (Dubberly et al., 2009).

We construe cybernetics as a process for understanding (von Glasersfeld, 1995) as well as a practice for operating in the world that focuses on systems that contain loops that enable the attaining of goals (Pickering, 2014). The term *cybernetics* comes from Greek roots meaning to pilot or to steer; on moving into Latin it becomes to govern. Some erroneously construe cybernetics to be mechanical. Some even hear in the word *system* the march of jackboots—unthinking, mechanical control. What interests us is quite the opposite—the messy chaos of natural and social systems, which cybernetics can help us begin to understand. We believe there is huge range for variation and possibility while applying the cybernetic frame to designing objects, interactions, services, and more. We also believe it is a misunderstanding to construe cybernetics as requiring a reductive stance or focusing on engineering. Glanville himself makes the point that Norbert Wiener ought to have published his most famous book *Cybernetics: Communication and Control in the Animal and Machine* after he had published *The Human Use of Human Beings*—because the former left an imprint of cybernetics as

---

3. From 2002 through 2007, the authors co-taught the course "Introduction to Cybernetics and Design" at Stanford University in Terry Winograd's Human-Computer Interaction program. Pangaro taught a related course in the School of the Visual Arts Interaction Design MFA program in New York and brings these perspectives to his position as chair of the masters program in interaction design at the College for Creative Studies in Detroit from April 2015. Dubberly uses the materials in lectures and courses taught at Northeastern University, California College of the Arts. For details of the approach, see Dubberly & Pangaro, 2013.

engineering grounded in mathematics, while the latter explains cybernetics as "a way of thinking and a way of being in the world" (Glanville, 2014a). The flowering of cybernetics in the 1940s came from conversations among a vast range of world-experts from both the hard sciences and the social sciences, all of whom celebrated the field as uniquely focused on a new way of seeing systems (von Foerster et al, 1950-1957; Dubberly & Pangaro, 2015).

The Rationale for Second-order Cybernetics & Conversations for Design

The structure of our argument is:

- If design, why systems?
- If systems, why cybernetics?
- If cybernetics, why second-order cybernetics?
- If second-order cybernetics, why conversation?

We now traverse that path and offer rationale and implications.

If Design, Why Systems?

Many of today's design challenges are *complex problems*, where an appropriate formulation of the situation is neither already agreed-to nor easy to characterize. However, through conversations within a design team, an agreeable characterization may be defined (the *problem formulation*) and then tackled by defining actions to improve the situation (the *solution*).

The industrial era changed the nature of design from design-for-making (insofar as there were any explicit design steps before making) to design-for-manufacturing. Beginning in the 20th century, *design-for-systems* becomes necessary, as evidenced from World War II when operations research as a field of practice and cybernetics as a systems discipline arose (Hughes, 1998). As argued in-depth elsewhere (Dubberly, 2014; Forlizzi, 2013), designers of digital systems are faced with the challenges of *product-service ecologies*. (Later we will widen the scope beyond digital and see that design-for-systems still applies.) This new design challenge is often exemplified by the iPod, but everything the same could be said for any portable networked device. While the user interacts with a physical device, the hardware's software connects to a network of communication systems (Internet) and databases (music stores) and marketplaces (music for sale), which has relationships to other actors (social community members, artists) and related aftermarkets. The complications of this system of systems must not be exposed to a user; and the designer must know enough about the system-to-system relationships to produce an effective design. Hence, designers must be conversant with this end-to-end mesh of (sub)-systems in order to design for a tractable set of rich choices from which the user lives her experience.

The rise of design-for-systems has further consequences. Good *form-giving* is largely table stakes—necessary but not sufficient to ensure the success of new ventures. New value-creation has moved to the development of systems. The term *platform* is often invoked in reference to complex, distributed interactions of hardware and software, networks and users, transactions and markets, for which primary examples are Alibaba and Amazon; Facebook and Google; Apple and Samsung (Dubberly, 2014).[4]

Design for complex problems that bridge product-service ecologies requires new skills:

> Looking at a specific system, recognizing the underlying pattern, and describing the general pattern in terms of the specific system constitutes command of the vocabulary of systems, reading systems, and writing systems—that is, systems literacy (Dubberly, 2014, p. 7).

If Systems, Why Cybernetics?

> One of the things I should do is try to make a little difference between cybernetics and systems, or see if there is one. (Glanville, 2014c, 2'28")

From the 1960s, The Club of Rome (Meadows, Meadows, Randers, & Behrens, 1972) popularized *systems dynamics* (SD) as a modeling language for complex systems, and since then Donella Meadows' and others' work has brought SD to a wide range of populations, including design students (Meadows & Wright, 2008). Conceived as a toolkit for explaining ecologies and economies, the vocabulary of SD—resource *stocks* and their *flows*—is well suited to its original application. However, we see limitations in SD for modeling systems for interaction. Meadows only briefly mentions regulation. SD does not clearly differentiate system behaviors that are the result of variations in levels (stocks as well as flows) from system behaviors that are the result of feedback. Perhaps most limiting is SD's lack of distinction between the effects of changes of levels (for example, an increase in population) and a deliberate act to effect an outcome. Goals require agency, and agency implies actions taken by participants that are based on data interpreted as feedback to the system's goals.

Goals and information are about the immaterial aspects of systems while stocks and flows are very much the materiality of them. The originators of cybernetics sought to make a clear distinction between the material and the immaterial. Ashby goes so far as to say "the materiality is irrelevant" (Ashby, 1956, p. 1) in order to further distinguish cybernetics as a discipline focused on information in purposive systems. As Glanville states while invoking Ashby, cybernetic systems are "not subject to the laws of physics and energetics, but subject to the laws of information, of messages" (Glanville, 2014a, p. 4).

---

4. The platforms mentioned are grounded in digital technology and therefore incorporate hardware/software infrastructure, but not all platforms are digital (see later example of the Schiphol Airport signage system). Our definition of *platform* includes the capacity for others to build systems within it, no matter the medium. We distinguish three levels of design: 1) design of *things to be used,* 2) design of *tools that can be used to make other things*, and 3) the design of situations in which others can create, that is, the design of platforms.

Because design involves human beings—what we want and how we might act to get what we want—systems literacy for designers must go beyond SD and incorporate goals and agency. Designers must therefore understand the workings of systems with agency. Cybernetics offers both language and models for understanding and describing such systems.

A cybernetic viewpoint on design also invites (if not demands) consideration of the capacity of a given system to achieve goals (whether imbued by a designer or inherent in the system itself). This of course is the concept of variety (Ashby, 1956). When the system is a team of designers, the question need be asked: Do we have the requisite variety to successfully design and construct an outcome that will achieve our goals?[5] This question raises other questions, how do these goals arise, and whose are they? To answer requires a shift to second-order.

If Cybernetics, Why Second-order Cybernetics?

> I have also developed the analogy between second-order Cybernetics and design so as to give mutual reinforcement to both. Design is the action; second-order Cybernetics is the explanation (Glanville, 2009, p. 22).

Today's most critical design challenges are global in scale and have direct impact on quality of life—and its very existence. They include the future of the climate, water, food, population, health, and social justice. They are characterized as *wicked problems* (Rittel &Webber, 1973) because the challenge to be addressed appears irredeemable. Even defining "the problem" is itself elusive, subjective, and controversial. Calling these situations problems is misleading; a better term might be *mess* or *tangle*.

It gets even worse. Wicked situations are impossible to solve fully; rather, we work as hard as we can to minimize their negative effects, but we cannot eradicate them. In part this is because these situations operate across complex systems of systems, with emergent and unpredictable behaviors, including *unintended consequences*, even when well-intended actions are taken. Now add that some of the systems employed are human networks, comprising ecologies of language and conversation, with concomitant ambiguity, conflict, and human defects at play.

In sum, creating a formulation of a wicked situation such that actions may be identified, whose execution has some likelihood of effectiveness, is a design challenge of the greatest degree of difficulty and greatest importance for our future.

Rather than speaking of solving in the context of wicked situations, the convention is to speak of positive change as taming. Taming wicked situations requires the acknowledgment of the need for *framing*—the subjective look at situations from a perspective that is only one possibility of many. The value of one frame above another is guidance to an effective path forward, usually through a frame's power to explain why the system behaves as it appears to. This is a form of

---

5. For elaboration of design for variety, which is beyond the scope of this paper, see Geoghegan and Pangaro, 2004.

taming complexity through language (von Foerster, 1984). Framing must support objective facts but only by being explicit about the values that forefront some "facts" above others. Fundamentally, it must create an argument for some design approaches above others—the design rationale. Systems dynamics and even first-order cybernetics are not enough:

> The systems-approach "of the first generation" is inadequate for dealing with wicked problems. Approaches of the "second generation" should be based on a model of planning as an argumentative process in the course of which an image of the problem and of the solution emerges gradually among the participants, as a product of incessant judgment, subjected to critical argument. (Rittel & Webber, 1973, p. 162)

Rittel is important in part because he is among the first to frame *design as politics*—as discussion and argumentation—as opposed to *design as art* or *design as science* (Simon, 1969). Similarly, Buchanan (1985) later framed *design as a branch of rhetoric*.[6]

Rittel points out that the stance of designer as expert problem-solver is largely a myth. There are few design problems with clear solutions. Design is not objective; it's subjective. It's messy. The designer never stands outside the situation. The designer is always part of the situation—and other constituents of the situation also have necessary roles to play in the design process.

Thus design becomes centered in an argumentative process that involves "incessant judgment, subjected to critical argument" (Rittel & Webber, 1973, p. 162). Rather than existing outside the design situation, judgment and argument appear inside when the stance is that of second-order cybernetics. For the shift from first-order to second-order occurs when the observer—the designer, the modeler, the problem-framer, the participant in design conversations—is aware of her observing.

In sum, design for wicked problems, and the required (re)framing, calls for second-order cybernetics, which makes the role of the observer explicit, which in turn makes explicit the subjective position of every design rationale.

---

6. There can be no mistaking that this approach to design has little to do with engineering qua problem-solving. Following Rittel and Buchanan, we situate design squarely in the realm of rhetoric. This does not, however, deprecate the value of rigorous modeling of systems nor the making of tools (for example, software and services). Software and services can be difficult to see—unfolding over time and space, intangible, often hidden or veiled. Absent clear referents (designations of the subject), conversations (and conversants) can become confused. Susan Star (Star & Griesemer, 1989) suggests the importance of *boundary objects* in supporting conversations between disciplines, by providing referents. Architectural plans, elevations, and all the rest of the architect's devices are boundary objects aiding conversations. (They are quite literally designations.) A traditional architecture education introduces these devices, starting with orthographic projection and moving on to isometric projection, perspective, and the rest. These constructions are a sort of language of their own, an argot of the profession. Software and service design is just beginning to develop such devices (its own forms of designation). Systems theory (e.g., systems dynamics, cybernetics, and the rest) offer distinctions and frameworks—a language—which designers can learn and use to create boundary objects, which can facilitate conversations about software and services (and their users, context, and environment) in the same way that plans, elevations, and sections facilitate conversations about buildings.

If Second-order Cybernetics, Why Conversation?

Conversation is the bridge between cybernetics and design (Glanville, 2014a, p. 8)

Design is a circular, conversational process (Glanville, 2009, p. 22)

Developing judgment and making arguments are, of course, forms of conversation. Glanville further tightens his assertion about the relationship of design and conversation by stating that conversation is a requirement for design, even when the conversation is with oneself, perhaps just using pencil and paper. (Schön, 1983, makes a similar point.) There is the person who draws and the (other) person who looks. The difference between these personae—between *marking* and *viewing*—is, in and of itself, a major source of novelty, Glanville claims. (We prefer the terms *variation* or *invention*. Our position on the role of novelty in design is given below.) Engaging multiple perspectives is a necessary condition for conversation, and without conversation, he writes, "You're not doing design, you're doing problem-solving."[7] Design, instead, is "to do something magical" and "to find 'the new'" (Glanville, 2014a, p. 10).

We state elsewhere (Dubberly & Pangaro, 2009) that conversational interaction is required in order to converge on shared goals. To share goals is to agree on (re)framing a situation in order to act together. We see the development of arguments in the course of designing (for or against different ways of framing situations) and the derivation of different choices or actions as the same as conversation. Thus we concur with Glanville's eloquent, albeit general, statements about conversation, cybernetics, and design.

However, we find some of Glanville's stated positions to be assertions without an accompanying rationale. For example, he was clear and even adamant that design knowledge is tacit, not explicit. We take this as part of his argument that design knowledge exists only in relation to action. If design is conversation, however, and if conversation is learning—very often, or at least consistently so in relation to design—then is not both the goal and the effect of the design conversation to make its subject explicit? We assert that for the major design challenges of today, making design knowledge explicit is a necessity. Form-givers may have the luxury of working alone, but designing systems and designing platforms require teams—and thus goals and methods must be made more explicit so that designs are coherent and actions are coordinated. Just as design is different than problem-solving, making choices in designing is different than making choices in creating a work of art. When designing, fit-to-purpose is the rationale for one choice above another; the question, of course, is do we agree on the purpose? When designing for systems, articulating that rationale is an irreplaceable component of the design conversation that takes place across the individuals, disciplines, and languages that comprise a design team.

---

7. While we accept the distinction between design and problem-solving, we can imagine typical cases of problem-solving that require conversation. For example, a team might discuss how best to break down a problem into more manageable components.

A retort might be that a given design conversation is about some specific situation or artifact—not about design. But then, a design conversation about design must be the subject of design education, and we arrive at the same point—making the tacit explicit is a requirement for effective design. Not doing so leaves design stuck in its medieval master-apprentice craft tradition, where change is slow, and innovation is difficult.

Implications for Designers

We have argued that 21st-century design requires conversation, as well that (in complete alliance with Glanville) design is conversation. When we say *conversation* we mean it explicitly in the second-order sense of recognizing our (subjective) participation in the process of framing and justifying our choices, and therefore our responsibility for it all.

If designers are to be responsible for the process of design, we must seek the most effective tools and methodologies—and to document, evolve, and disseminate them into the community of design and into the world of wicked problems.

Therefore, designers must themselves be responsible for systems literacy as the foundation for design; for working within a second-order epistemology where they take responsibility for their viewpoints; for processes of collaboration through conversation; and for articulating their rationale as an integral part of their process. This has deep implications for the development of curricula for teaching design.

Implications for Teaching Design

Glanville was influenced by his experience of design methods during his time as a student at the Architectural Association in the 1960s. Perhaps it was in rejection to prescriptive design methods of the first generation that he came to prefer to say that design is "at once mysterious and ambiguous" (Glanville, 2014b, pers. comm.).

We agree that when narrowly interpreted in its first-order form, cybernetics as engineering, may suggest a sort of problem-solving which accepts or even assumes goals rather than inviting conversation about what our goals should be. But in its second-order form—with subjectivity, values, and responsibility explicit—isn't teaching design as cybernetics more common-sense than straight-jacketed engineering, more about possibility than determinism, more emergent than mechanical? Teaching vocabulary and grammar does not deny poetry. Quite the contrary: A knowledge of vocabulary and grammar, if not a prerequisite, seems at least a more fertile ground for the emergence of poetry, and her sister, delight.

Novelty, Design, and Second-order Design

> For me, one of the most important things is how to find novelty, and that I don't think can be done by specification or purposeful action, it needs wobbly conversation and deep speculation. After it's found, it can be specified. (Glanville, 2014b, pers. comm.)

While not presuming too much about Glanville's possible elaborations on the relationship of novelty and design, we want to be clear about ours: Novelty is not the primary goal of design. (There is a risk that traditional designers will hear the pursuit of novelty as the pursuit of new form for its own sake.) Like Glanville, we embrace conversations for design, specifically as a way of discovering new goals and new opportunities, as we co-construct our shared frames and persuading arguments. But as yet tacit in our argument is the role of value and values. Design is a particular set of conversations which explicitly and implicitly, whether to oneself alone or with others, embody what we value and what we seek to conserve. Maturana's framing of possible change in the context of what we do not wish to change is directly useful and actionable:

> Every time a set of elements begins to conserve certain relationships, it opens space for everything to change around the relationships that are conserved. (Maturana et al., 2013, p. 77).

Of course we must be aware of what we are conserving, to open the possibility of second-order change. Unstated but what we hear implied in Glanville's position, is the notion that the results of design should not be fixed—that is, that designers create possibilities for others to have conversations, to learn, and to act. This idea may be the most important of all. It represents a paradigm shift. Le Corbusier's publication of *Le Modulor* may be a fulcrum point, the visible signal of the new paradigm. (Though moveable type with its inherent reuse sets the stage for what comes after modernism, even as moveable type creates the revolution of modernism itself.) To single out one example, the Schiphol Airport signage system from 1967 by the Dutch firm Total Design and Benno Wissing is one of the first and most famous examples in practice—creating not a complete system, but a system in which others can create. As a platform for creating—in our terms, a platform for conversations for designing—a signage system is quite limited, but still the outlines are there. The relationship of designer to outcome is changed: The signage system is never completely finished, never completely specified, never completely imagined. It is forever open. Second-order design is born.

We see this as the emergent space of design for the 21st century and aim for it as our goal. Whether designing interactive environments as computational extensions of human agency or new social discourses for governing social change, the goal of second-order design is to facilitate the emergence of conditions in which others can design—and thus to increase the number of choices open to all.

References

Ashby, W. R. (1956). *An introduction to cybernetics*. New York: Wiley.

Buchanan, R. (1985). Declaration by design: Rhetoric, argument, and demonstration in design practice. *Design Issues, 2*(1), 4-22. (Reprinted in V. Margolin, Ed., *Design dscourse: History, theory, criticism*, The University of Chicago Press, 1989)

Dubberly, H. (2014). *The networked platform revolution*. Presentation given at IIT Institute for Design, Chicago. See http://presentations.dubberly.com//ID_Networked_Platform.pdf.

Dubberly, H., Chung, J., Evenson, S., & Pangaro, P. (2009). *A model of the creative process.* Report commissioned by Institute for the Creative Process, Alberta College of Art Design, Calgary, Canada. See http://www.dubberly.com/concept-maps/creative-process.html.

Dubberly, H., & Pangaro, P. (2007). Cybernetics and Service-Craft: Language for Behavior-Focused Design, Kybernetes, Volume 36 Nos. 9/10. Seehttp://www.dubberly.com/articles/cybernetics-and-service-craft.html.

Dubberly, H., & Pangaro, P. (2009). What is conversation? How do we design for effective conversation?, Interactions Magazine, publication of the ACM, July/August 2009. See http://www.dubberly.com/articles/what-is-conversation.html.

Dubberly, H., & Pangaro, P. (2013). Course description for "Introduction to Cybernetics and Systems for Design." See https://sva.instructure.com/courses/181509.

Dubberly, H. & Pangaro, P. (2015). How cybernetics connects computing, counterculture, and design. In *Hippie modernism: The struggle for utopia* (pp. 126-141). Minneapolis, MN: Walker Art Center.

Geoghegan, M., & Pangaro, P. (Eds.). (2009). Design for a self-regenerating organization. *International Journal of General Systems, 38*(2). (Paper originally presented at the Ashby Centenary Conference, March 4-6, 2004 University of Illinois, Urbana)

Glanville, R. (2002). Second-order cybernetics. Invited chapter for *Encyclopedia of Life Support Systems*, UNESCO. Retrieved March 3, 2014 from www.eolss.net/ . (Reprinted as chapter 1.11 of Glanville, R., 2012, *Cybernetic Circles, The black bσx, Vol. 1*, echoraum-WISDOM, Vienna.)

Glanville, R. (2014a). *How design and cybernetics reflect each other.* Paper submitted to Relating Systems Thinking and Design RSD3 2014 symposium, "Relating Systems Thinking & Design 3." Retrieved November 6, 2015 from http://systemic-design.net/wp-content/uploads/2014/08/Ranulph_Glanville.pdf

Glanville, R. (2014b). Email correspondence with the authors.

Glanville, R. (2014c). Audio of keynote presentation at Relating Systems Thinking and Design RSD3 2014 symposium. Retrieved from https://www.youtube.com/watch?v=tTN_9mJIWNw&feature=youtu.be on November 6, 2015.

Hughes, T. (1998). *Rescuing Prometheus.* New York: Vintage Books.

Maturana, H., & Dávila, H. (2013). Systemic and meta-systemic laws. *Interactions Magazine, XX*(3; May/June).

Meadows, D. H., Meadows, D. L., Randers, J., & Behrens W. W., III. (1972). *Limits to growth.* New York: New American Library.

Meadows, D. H., & Wright, D. (2008). *Thinking in systems: A primer.* White River Junction, VT: Chelsea Green Publishing.

Pickering, A. (2014). *The next Macy Conference: A new synthesis.* Keynote for IEEE Conference Norbert Wiener in the 21st Century, Boston, June 2014.

Rittel, H. W. J., & Webber, M. M. (1973). Dilemmas in a general theory of planning. *Policy Sciences, 4,* 155-169.

Schön, D. (1983). *The reflective practitioner.* New York: Basic Books.

Simon, H. (1969). *The sciences of the artificial.* Cambridge, MA: The MIT Press.

Star, S., & Griesemer, J. (1989). Institutional ecology, 'translations' and boundary objects: Amateurs and professionals in Berkeley's Museum of Vertebrate Zoology, 1907-39. *Social Studies of Science, 19*(3), 387–420.

von Foerster, H., et al., (Eds.). (1950–1957). *Cybernetics: Circular causal and feedback mechanisms in biological and social systems: Conference transactions.* New York: Josiah Macy, Jr. Foundation.

von Foerster, H. (1984). Disorder/order: Discovery or invention? In P. Livingston (Ed.), *Disorder and order: Proceedings of the Stanford Interaction Symposium* (pp. 177-189). Saratoga, CA: Anima Libri. (Reprinted in von Foerster, H. (2003). *Understanding understanding: Essays on cybernetics and cognition* (pp. 287-304). New York: Springer.

von Glasersfeld, E. (1987). An introduction to radical constructivism. In *The construction of knowledge* (Chpt 10; pp. 194-219). Salinas, CA: Intersystems Publications.

von Glasersfeld, E. (1995). *Radical constructivism: A way of knowing and learning.* Studies in Mathematics Education Series: 6. London: Falmer Press.

Bunnell, P. (2014). *Straight and Curved.* Un-retouched photograph.

Cybernetics and Human Knowing. Vol. 22 (2015), nos. 2-3, pp. 83-87

Designing Exploring as a Second-order Process:
A Legacy of Ranulph Glanville

Robert J. Martin[1]

The Ranulph Glanville I knew was concerned with change, especially change within individuals and groups that comes about through a circular process of conversing and reflecting that leads to new understandings that lead to new ways of experiencing that lead to conversing and reflecting; in other words, a process of acting that requires reflective practice (Schön, 1983). One approach to reflective practice (and the practice Ranulph claimed as his own) is designing. I have come to see his work as a process of designing the activity of exploring, or, simply, designing exploring. Ranulph characterizes designers (and design) as follows:

> Designers work in a way ... that allows them to deal with very complex, ill-defined and ambiguous situations that would probably be inaccessible using conventional approaches. The outcomes of design are generally novel but can never be argued to be the best solution ... Design is necessarily a constructivist, second-order cybernetic activity. (Glanville, 2008, p. 75)

Designing is a constructivist activity in that those involved in designing are aware that they are making, not finding, what they do. Designing is a second-order cybernetics activity in that it is an intentionally circular process operating on itself. In the case of designing exploring, it is also a process whose product cannot be fully or, at times, even adequately known, as we will see in the following paragraphs.

Composing Exploring

It isn't necessary to talk about cybernetics to do cybernetics, and it isn't necessary to talk about second-order processes in order to carry them out. The power of designing as a second-order process is that it can be applied to many situations. One way I got to know Ranulph was through our practice as composers of music. Ranulph was a serious experimental composer whose interests were in exploring constraints through electronic music. This was not a side interest but an integral part of how he practiced second-order processes. His characterization of designers and design (above) is also an excellent definition of composing as an experimental process. In the following paraphrase of the paragraph quoted above, the word *composers* replaces the word *designers*; and the word *composing* replaces the word *design*. All words not in brackets are Ranulph's:

---

1. Email: rmartin@truman.edu

[Composers] work in ways (conversations with self via media such as pencil and paper, [computer programs, musical instruments or other sound sources]) that allow them to deal with very complex, ill-defined solutions. The outcome of [composing] is generally novel, but can never be argued to be the best solution. (Glanville, 2008, p. 75)

An important point (and a point linked to second-order cybernetics) about art, including music, is that art is not about creating beautiful objects as such, but is about solving problems—and, in the process, coming up with (sometimes) something new.

Reluctant to push his own music, Ranulph nevertheless agreed to present a video piece of his entitled "Generator" at an ASC Convention held at Ryerson University in Toronto. In conversation, Ranulph described the piece as monolithic, meaning that it was based on one sound and one image. For approximately twenty minutes, the viewer sees a grainy video of a long hallway with a noisy machine (some sort of ventilator) providing the sound. The machine, a *found* noise generator, retains its identity over time while changing on a micro level both from moment to moment due to the mechanical nature of the sound source and also transforming ever so slowly as a result of having been run through several software processes.

Ranulph's interest was in taking the video camera to the edge of what it was able to do and then seeing what resulted. He wasn't interested in using the technology of the camera and the software to make music; he was interested in exploring the limits of technologies in ways they weren't designed to be used; the result was the piece. Of course the result is nothing that resembles a typical piece of music, but I found it fascinating and had no trouble listening and watching for the full duration of the piece. Ranulph allowed me to use this piece over a number of semesters in an undergraduate interdisciplinary course I taught at Truman State University, Creativity in the Arts and Sciences. Students liked it, hated it, or endured it but were otherwise indifferent. This diversity was perfect from my standpoint: the diversity of opinions meant that we could explore our reactions to Ranulph's exploring; everyone could come to their own conclusions while having to consider opinions different from their own—including an opinion about what could be regarded as music implicit in the piece itself. Both processes (listening and then conversing) brought students to question their own understandings and, through the conflict between their views of music and the piece, and between their views and the views of others, to enlarge their understandings—a second-order process as I saw it.

Designing Exploring as a Participative Process

Ranulph's papers and lectures (some of which are available on YouTube) are designed to include readers or listeners in exploring a set of ideas that require those readers or listeners to engage in thinking about their own process of understanding or acting—which is a second-order process. The implicit goal was not to inform readers and listeners about second-order processes but to involve them in actively questioning their own views and understandings, and thus learning about a second-order process by participating in it.

Taking this process of engaging participants in second-order processes a step further, Ranulph worked to design a series of conferences that would engage participants more directly in conversing and reflecting on a topic or problem. As part of a group collaboratively engaged in designing one of these conferences, I can attest to the fact that the nature of such designing, including handling all the details, is "very complex, ill-defined and ambiguous" (Glanville, 2008, p. 75). The experience of participating in the conversational conferences was challenging and sometimes frustrating, both for the designers and for the participants. Again, I did not always like the ambiguity and uncertainty, let alone the conflict and drudgery that is necessary to address a topic or problem where everyone has different language to talk about the same words (a word being the name of what it evokes in each of us, not the name of something per se); but I am choosing to embrace it. It is important to mention conflict because, while some might interpret the mention of conflict as a criticism of the designers and therefore something to be avoided, this elephant in the room is a necessary and inevitable part of designing a process that departs significantly from the process of other conferences. If we want the creativity that comes from second-order designing, we need to embrace—or at least be willing to live with—conflict. I do not (and did not) find this easy. Now, from a distance of some months, I see that we need to do more of such designing.

Understanding Understanding and Participative Process

In summary, what I have described in this brief paper are Ranulph's efforts to bring us to a place where we can engage with him or with each other in second-order processes. We can then decide whether we wish to continue.

When second-order cybernetics became a constructivist undertaking, it moved away from the study of control as an objective possibility. After all, if systems consist of descriptions rather than objective realities, we can never be sure that we can intervene in ways that are predictable and desirable. This doesn't mean that we shouldn't act; it does mean that we might want to approach acting in regard to systems—including ourselves, one another, and nature (including the planet)—with humility, circumspection, and a light touch.

What direction do we now wish to take? One direction is to embrace the ambiguous and uncertain path of designing exploring (versus or in addition to our traditional and often very satisfying ways of doing science). Ranulph's path as a writer, speaker, composer, collaborator, member of many societies, as well as president of the ASC, was to embrace second-order processes, including applying cybernetics to itself. Applying cybernetics to itself is one of our favorite and best mantras. Do we make it more than a slogan—and how do we make it more than a slogan? I have presented Ranulph's approach to engage us in this process as a way of encouraging us to reflect on this part of his legacy—and then to act.

At all scales, from the personal to the planetary scale, we face "very complex, ill-defined and ambiguous situations that would probably be inaccessible using conventional approaches" (Glanville, 2008, p. 75). This is a moment where we, individually and as members of various groups, have the opportunity to decide in what direction we want to take second-order cybernetics and systems science. Our choice is to embrace the ambiguity and uncertainty of designing our own exploring—or to remain as we are. I choose being involved, and, as a consequence, uncomfortable and uncertain. Only then can I support others who are doing the same or more than I am.

Ranulph was pessimistic about the future of the ASC precisely because of the problem of getting others involved—and he put in innumerable hours over a decade to keep the ASC going. As a member of the current executive board, I see how much my colleagues do in designing and bringing about the journals, websites, and conferences that make our participation as writers, speakers, and conversation makers possible.

Many of us want to create change in ourselves and others as a way of moving toward a future where life is both rich and sustainable. For better or worse, change happens—mostly without our understanding, though almost never without our efforts. The possibility of increasing our understanding of understanding (a second-order process)—including how understanding changes thinking and acting can be a source of motivation for continuing to experiment with designing opportunities for second-order thinking (such as conversational conferences).

At the 50th anniversary conference of the ASC in 2014, Michael Hohl presented a simple drawing that summarized our challenge: The viewer sees two white circles against a black background. A small gulf separates the two circles. The smaller circle is labeled "your comfort zone"; the larger circle, using a third of the available space, is labeled "where the magic happens" (Herrick, 2012). A copy of the drawing is on my desk lamp as I write this. The two circles and seven words do not tell us what to think or do; they invite us to contemplate our own existential situation—the paradox that the good stuff always seems to lie outside our comfort zone. Ranulph brought his second-order designing theory to bear on what he practiced as a writer, speaker, conference planner, and conversation maker, and, by stepping out of his comfort zone, he invited others to step out of their comfort zones.

Designing, especially a second-order process like designing exploring, means getting our hands dirty—putting in the time required to plan and carry out projects—such as conversational (or other types of) conferences or other endeavors that bring theory and practice together in a circular reflective process. I resist stepping out of my comfort zone, and I know I want to keep stepping out of my comfort zone. I also want to be part of a community where others step out of their comfort zone.

References

Glanville, R. (2008). Designing complexity. *Performance Improvement Quarterly, 20*(2), 75-96. Retrieved September 22, 2015 from http://onlinelibrary.wiley.com/ (doi/10.1111/j.1937-8327.2007.tb00442.x/abstract)

Forester, H. von. (Ed.). (1955). *Cybernetics: Circular causal and feedback mechanisms and social systems: Transactions of the Tenth Conference, April 22-24, 1953, Princeton, N.J.* New York: Josiah Macy, Jr. Foundation.

Foerster, H. von & Broecker, M. S. (2010). *Part of the world: Fractals of ethics—A drama in three acts* (B. Anger-Daiz, Trans.). Heidelberg, Germany: Carl-Aurer-Systeme Verlag.

Herrick, N. (Ed.). (2012, April 1). *Where the magic happens ...* Retrieved September 22, 2015 from http://kidsstylehub.blogspot.de/2012/04/where-magic-happens.html

Schön, D. A. (1983). *The reflective practitioner: How professionals think in action.* New York, NY: Basic Books.

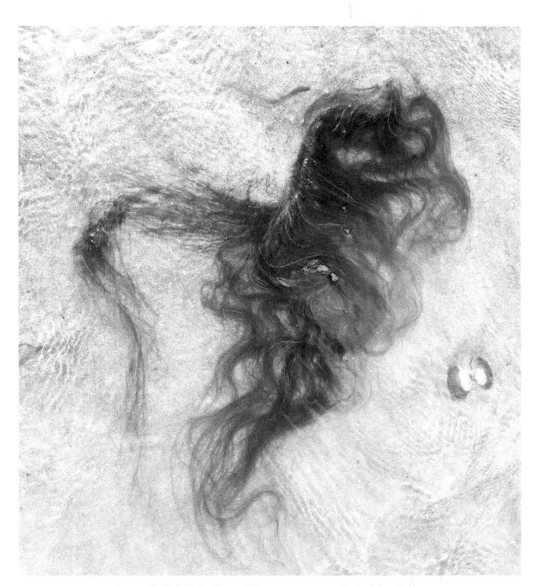

Bunnell, P. (2015). *Green Wig, Lost.* Un-retouched photograph.

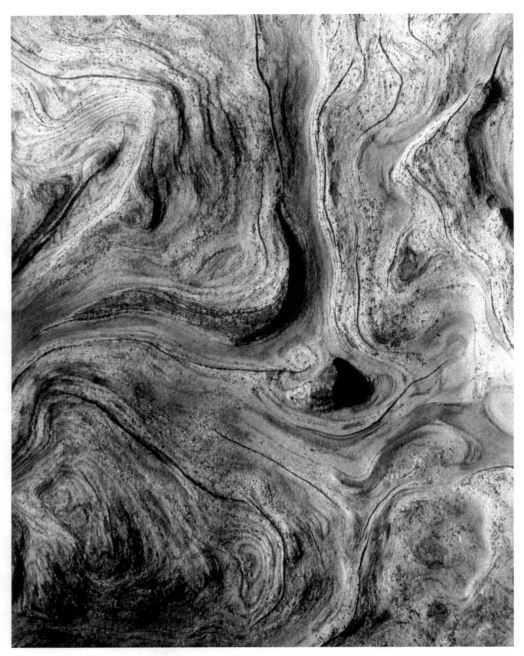
Bunnell, P. (2007). *Choices*. Un-retouched photograph.

Cybernetics and Human Knowing. Vol. 22 (2015), nos. 2-3, pp. 89-97

Preferences in Architectural Research

Gerard de Zeeuw[1] and Rolf Hughes[2]

Academics often claim to have insufficient time to do research, and yet academic life provides many freedoms and opportunities. It is possible to talk with other staff without having to support the "business"; it is possible to meet and share discussions that started many years, even decades, ago. It certainly was possible to do so in 1974 when one of us met Ranulph Glanville for the first time. His 20-minute presentation stimulated a 15-minute question, possibly to the despair of the chair of the session. His ideas on objects were highly stimulating.

This event started a trip to the bar as well as life long discussions. We also started to participate in each other's work—as when Ranulph spent some years as staff in the Centre for Innovation and Cooperative Technology of the University of Amsterdam. He made strong and important contributions to the series of conferences called "Problems of ... ." Among them was the way he was able to meet "low flying pubs" and still be present the next day to be a model for aspiring academics. He opened their world. Later Ranulph helped both of us to participate in the exciting developments linking research and architecture in Belgium.

Here we continue to discuss some parts of his work as a tribute to a rare mind—a person who saw the present as the beginning of the future rather than as the end of the past.

Keywords: sets of observations, improving preferences, observational research, non-observational research, constraint.

Introduction

The authors didn't know each other when, in 2006, they were invited to help develop the Research Training Sessions (RTS) of the doctoral programme at the Hoge School Sint Lucas.[3] We were coupled from on high to support participants becoming researchers. Our backgrounds differed, in nearly everything—except, clearly, in our interest in research (Hughes, 2012). We also differed from our students, as nearly all of them were working architects. The difference deterred neither them nor us. Together we happily discussed the highs and lows of research and architecture. While the students developed the plans for their studies, we did something similar. We tried to provide a conceptual answer to the question how to combine research and architectural practice.

Part of the reason we were invited may have been that we had already attempted to do research in areas such as art, design, social innovation, management as well as architecture. All areas are characterized by having thrived on their own steam for a

---

1. Gerard de Zeeuw is emeritus professor of mathematical modelling of innovation (University of Amsterdam), of management research (Lincoln, UK) and of design research (Leuven, Belgium).
    Email: gerard@cict.demon.co.uk
2. Rolf Hughes is head of research and professor of artistic research at the newly created (2014) Stockholm University of the Arts, as well as elected Vice President of the International Society for Artistic Research.
    Email: rolf.hughes@uniarts.se
3. It has since become the Faculty of Architecture, Catholic University of Leuven.

long time so there are many wonderful exemplars—but mostly without the benefit of research. Even so there are the gods, for example of architecture: people who have had a strong impact on our built environment, based on their individual creativity and on thousands of years of building experience. We report here on the task we set ourselves: to explore and, if possible, extend what a combination of research and architecture might entail, in cooperation with the members of the RTS.

Values

Professional expertise seems to derive mainly from the human ability to talk about our experiences and thereby change them. That ability is a miracle itself if one considers that it is only 35000-40000 years ago that we (and possibly the Neanderthals) started to draw or talk via pictures, that is, via their observations of horses, mammoths and themselves, as the caves of France and Germany demonstrate (Walter, 2015). Although we cannot claim that people other than the artists understood the drawings, their ubiquity suggests that they provided a crucial message. Many years later the ancient Greeks took another vital step when they tried to ascertain the importance or value of such talking. Did the drawings depict correctly or did they help only to set aside and value some experiences?

It took some further millennia for people to realize that focusing on observations made certain forms of talking more effective. Only after he had enhanced his observations by the use of a telescope, did Galileo (1610/1989) succeed in convincing (some of) his contemporaries that the moons of Jupiter do not circle the earth. Descartes (1637/2005) argued that Galileo's success depended on distinguishing what we observe (clearly and distinctly) from how we talk. Popper (1959/2002) showed how individuals may emulate successes as well as that searching for high quality never stops. Lakatos (1978) and Kuhn (1962) clarified that stopping requires that individuals talk to others and thereby link parts of their observations (so they recognize that a small light near Jupiter is a moon).

In contrast, when architects talk they tend to focus on preferences, that is, on what they or others want. Do buildings need to have balconies—always or only in cities? The difficulty architects face is that one cannot combine preferences into a preference that all accept as shared and in that sense has high quality. The concept of quality is a strongly contested one, therefore, in architecture as well as in art.[4] As indicated, researchers try to resolve a different difficulty, that is, they focus on combining observations into an observation that all can share. In the past four centuries they have proved able to identify such high quality observations in a number of areas. This difference leaves little room for architects to share in the researchers' tradition.

Why would they wish to do so, therefore? Is it because architects perceive such sharing as having been successful elsewhere? Or would there be space to do the reverse and have researchers share in the architects' tradition to help study

---

4. To emphasize this difference one may paraphrase a famed saying: there is no accounting for differences in taste, but there is for differences in observations. See also footnote 7.

preferences? We decided to concentrate on the second question. Although this meant going against mainstream attempts (Hughes, 2014), we realized that we were in very good company. Others went before us, for example philosophers like Heidegger (1929) and Wittgenstein (1986), or researchers like Von Foerster (1995), but also and most importantly in the present context, Ranulph Glanville. By wanting to study preferences, he wanted architects to inherit the power of research.

Ranulph sometimes referred to the time he started to appreciate *models for* as distinct from the usual *models of*.[5] It seems to have been an epiphany. We provided its source, as he graciously acknowledges (Glanville, 2005, p. 87). Although we shared the experience, we deviated since. This makes it attractive to try to put his work into perspective, partly by comparing it with our own efforts. This way we are able to celebrate its achievements, but also identify where there is a need to explore further. We argue that his strict separation between observing and talking is too limiting. Observations are modified by the way we talk and vice versa. This means that we should not only determine the value of the way we talk, but also how choosing that way may increase its value.

Observational Research

We take Glanville's (2002, p. 82; 2004) notion of research as our reference point. He suggests that researchers aim to link individual reports of observed patterns to sentences so that they can replace the reports. Such replacement would make the link unique and of high quality (Suppes & Zinnes, 1963). It can be found if one has to solve for one unknown given two knowns—but not otherwise, that is, in case of experiences other than observations. The first solution is sequential, that is, one identifies a set of observations, then the sentences and finally the link; the second is simultaneous, that is, one modifies the set as well as the sentences until a high quality link is found. Both solutions constitute formal models for researchers to identify a model of the set.

Examples include fitting a curve to a set of observations or averaging those in the set as a best estimate of the observation of some "real" event. These sentences (curves and averages) constitute improved (reports of) observations, that is, observations better than those in the set. Their links obviously cannot be found, however, if the set of observations and the sentences depend on each other and hence do not provide two knowns, but only one. This situation may occur when one allows a preference to determine the set. In this case only one known remains, that is, the preference. Such situations, including when the set contains preferences as well, are covered by Arrow's (1950) impossibility theorem. It states that the desired link cannot be found.[6]

---

5. The etymology of the word *model* includes the Latin *modellus* (diminutive of mode), which refers to a prototype: it is a model of the person whose statue is intended, and it is a model for the sculptor to create the final statue. See also Hughes and Monk (1998).
6. This impossibility is reflected in national sayings, including those of the ancient Romans. They claimed that "there is no accounting for taste" (or rather "de gustibus non disputandum est"). Differences may be resolved by voting, but also by luck or force, such as that of a dictator.

This means that a research procedure is only able to add value to observations—but not to experiences like preferences, objectives and emotions. As Arrow notes, the difficulty may be resolved by voting, that is, only temporarily. This seems to be precisely what architects do in everyday life. They tally clients' needs and decide on the result. Conflicts are resolved by client and/or architect acting as dictator, for example for aesthetic (the architect) or financial (the customer) reasons. Conflicts between architects and stakeholders seem to depend even more explicitly on voting—for example in urban design. According to Arrow, whenever they wish to deal with preferences, architects will find it hard to avoid voting.

Second-order Cybernetics

This difficulty was already identified quite early; for example in 1857 by Pasteur (see Latour, 1999, pp. 113-145). Resolving it requires that people develop as professionals—unless another way forward can be discovered. The interest in finding a research-connected approach grew quickly after WWII. Various solutions were explored, which led to subjects like operational research, decision-making, gaming and cybernetics. In each it is either attempted to constrain the set of preferences or the way of talking—to create two knowns again so the unknown link can be derived. In cybernetics, the constraint is a pre-selected preference (Stachowiak, 1969). It is implemented, for example, in the design of a thermostat, the set temperature of which users can modify and accept temporarily.

There are other solutions, for example that of mechanical design as part of game theory (Maskin, 2008). Already early in his career Ranulph similarly became aware of the difficulty, i.e. of the 'impossibility' of adding value to preferences. In his PhD he proposed a procedure to make the set of observations independent—so there would again be two unknowns (Glanville, 1975/1988). Later he contributed to the development of the area of second-order cybernetics. As Ranulph tells the story, it originated from Margaret Mead's suggestion during a conference to apply cybernetics to the group of cyberneticians present—possibly because they were a bit too disorderly (and didn't behave only as *observing systems*). She thought they should be able to steer themselves.

While this may have been a fleeting thought at the time, it led to an innovative development. If participants steer only themselves, their set of reported observations depends on their preference. To re-establish independence, participants will have to constrain their individual sets.[7] They may achieve this by accepting what others do as constraints on their own preferences. This may allow them to modify their preferences so they do not conflict with those of others. The result should allow those who aim to behave like cyberneticians (as their model for) to act as internal observers and develop improved models of themselves, as well as act as external observers and develop an improved model of what the interactions lead to.

---

7. Such sets include observations of what individuals contribute when they realize their own preference, given those of others.

Ranulph advocates this approach to help architects and researchers talk to each other. It involves adding at least one cybernetician to cooperate with a first so together they are able to collaborate in steering their group and avoid resisting each other. In this case there is no need to vote to resolve dependencies, therefore, or for a dictator—someone to tell others what preference to accept. Next he argues that a number of authors have found second order cybernetics useful in supporting architects. On a wider, strategic level he also notes that the approach fits Glasersfeld's (1984) radical constructivism, which thus might serve as an alternative way to design architectural research.

Although Ranulph repeatedly emphasized that second-order cybernetics was developed to improve preferences, how this is done is not made explicit. Having access to high quality observations is taken to be sufficient to realize any preference. In other words, to control a process one must understand its antecedents (as a model of). In practice there is much give and take, however. When observations are insufficient to realize existing preferences, either the observations are modified (via additional constraints, see above) or the preferences are. This means that convergence towards preferences that are temporarily mutually acceptable, if any, depends on achieving high quality internal and external observations.[8]

Convergence is assumed, therefore, it is not guaranteed. The situation seems similar to that of Adam Smith's *invisible hand*. Prices converge towards a price (not necessarily a single one) that is preferred, internally and externally—if the market (i.e., the conglomerate of interactions) is able to resist disturbances and sustain itself. The analogous process in second-order cybernetics similarly helps to bypass Arrow's impossibility. It is applied when observational research fails, that is, when it generates difficulties by attempting to study preferences as if they are observations.[9] Ranulph notes that certain concepts are analogous as well, like models for (rather than models of) and *knowing*, that is, the competence to act when constrained by interactions (rather than *knowledge*).

Non-observational Research

Like Ranulph, but based on our previous experiences as well as the discussions in the RTS meetings, we recognized our basic difficulty as finding ways to help researchers and architects talk and understand each other. We preferred to focus on the preferences, however, so as not to deal with them indirectly, that is, via internal and external observers. We asked how individual preferences might be linked to preferences that the individuals involved might accept, at least temporarily—thereby

---

8. It is tempting to envision what the interactions converge to as a collective (e.g., a market, see below), with its own preferences. This would undermine the argument, however. The convergence only refers to the way each participant starts to constrain what others contribute. The collective is an not part of the convergence and hence an epiphenomenon.

9. There appears to be little resistance to the idea that, if cyberneticians participate in the scientific enterprise, so will second-order cyberneticians: both solve Popper's demarcation problem (to do research rather than non-research). Improving preferences is not guaranteed.

making Arrow's impossible possible? To answer this question we also recognized the need to introduce constraints—just like the second-order cyberneticians. The most obvious are the ones that derive from the interaction with others.

Suppose, for example, that an individual wishes to build a house. He or she may try to design one that has high quality, as defined. A vote might help, but that would constitute a social arrangement rather than research. We preferred to initiate an interaction, therefore, starting with the commissioners and expanding to users and stakeholders. This way we would link individual preferences to each other so they constrain each other mutually.[10] This process thus will serve as a test: When the changes lead to a design being chosen, the preferences will have been improved. Eventually the interaction may become resistant to changes such as defections, new comers or dictators. When this is the case we consider the link to be of high quality.

This type of approach appears to resolve Popper's demarcation problem just as second-order cybernetics does. It also takes advantage of interactions among acting participants as a way to constrain sets of experiences so dependencies can be avoided. It also does not operate from a distance. It appears more focused, however, as the convergence towards resilient interactions is part of the approach, not something that depends on another approach, that of finding high quality observations. In addition, it can be said to constitute a *study of constraining*, that is, a search for models for individuals to constrain each other and thereby improve their preferences—rather than a search for improved internal and external observations.[11]

This study of constraining differs not only from second-order cybernetics, but also from other approaches that also aim to deal with preferences—such as voting. It differs in terms of its product as well as its process. Its product may refer, for example, to tangible things, such as a bicycle, a house, a piece of music, and so forth. Such things constrain an acting body (not that of a body at rest): People on a bicycle are able to travel faster than without it, but they lose most of their ability (e.g.) to swim. This transformation suggests an improvement. The quality involved is its precision in delineating what is changed. Bicyclists lose certain abilities,[12] but not others such as being able to choose a direction or to communicate—when on a bicycle (Roshchupkina, 2015).

The approach also differs in terms of its process. It does not refer to organizing participation (the participative approach)—and hence does not relate to processes that emphasize co-creation (Reason & Bradbury-Huang, 2013). The interactions are expected to develop in improving participants' preferences. There is also no intention

---

10. It may happen of course that additional as well as nested interactions are necessary to resolve the dependency issue—and hence that hierarchies of such interactions develop (Vahl, 1994).

11. We would not be able to say that, like second-order cyberneticians, "everything said is said by an observer." We would prefer "everything said is said by an actor or narrator."

12. Participants develop a competence, as indicated, i.e. to use preferences as benchmarks to develop what is acceptable to individuals (Agar, 2013; Hughes, Anstey and Grillner, 2007). The resulting competencies may survive beyond their temporary engagements. The competence to play fair, for example, developed when team sports became popular (Mangan, 1986). It became part of the British civil service culture. Later it was replaced by the preference (and competence) to compete.

to involve users of buildings, for example to help determine (as in a voting procedure) what faucets to install or what colors to introduce (Research Grants, 2012). The aim is to constrain activities (see above), but also to develop ways of interacting that can be or have been tested as to their power to engage people to act by realizing their preferences without encroaching on others' preferences.

There is another, more fundamental difference. We have called our approach non-observational to emphasize its special kind of input, namely, preferences (and other striving emotions) as well as a special kind of output, namely, ways to gain advantages when losing others—by initiating interactions. The emphasis on non-observational experiences is important and fundamental, as indicated, but also slightly misleading. It is important in that the aim is to improve preferences—in contrast to second-order cybernetics where the aim is to do so indirectly, that is, via internal and external observers. It is misleading in that observation still plays a role, namely, of checking whether interactions still change when new members start to interact.

Related Work

That difficulty we discussed during the RTS meetings was identified a long time ago ——although obviously in a different context than that of linking the way researchers and architects talk. It is no surprise that the literature already contains many attempts to improve intentions, objectives and preferences. They date back to Marx (1867/2013), but more specifically to Lewin (1948) who tried to liberate and improve the situation of individuals imprisoned in some form of interaction—by unfreezing, changing and refreezing that interaction. Others focused on ways to identify preferences that people would have to accept as guides: to optimize (operational research), to link to (Maskin, 2008) or achieve strategically (Von Neumann & Morgenstern, 1944/2000).

Systems research was similarly conceived to deal with the difficulty of the mutual dependence of sets of observations and sentences to talk about them (Von Bertalanffy, 1968; de Zeeuw, 2006). It stimulated the development of procedures to mobilize people's resources (e.g., Flood & Jackson, 1991; Rosenhead, 2005). While there is and was a clear interest in linking these procedures to research (Checkland & Holwell, 1998), the efforts of second-order cyberneticians appear more consistent and explicit. Our own efforts date back to the same period, that is, to work on methods and quality criteria to support interaction and conversation (Pask, 1975; de Zeeuw, 2007, 2010; Glanville, 2002; Axelrod, 1984; Rosen, 1991).

Observational research focuses on sentences, that is, on how we talk about observations. Its results constitute declarative knowledge. The alternative is to focus on procedures and to acquire procedural knowledge. The aim of the latter approach is to identify ways to instruct, guide, advise, coach people to improve their preferences. It also has been noted that declarative knowledge is what single individuals are able to acquire—while being guided by their interactions, that is, by their paradigm (Kuhn, 1962; Lakatos, 1978). This does not imply that non-observational research constitutes a new paradigm. Ranulph as well as we, interpret architectural research rather as being on the level of paradigms rather than as part of a single paradigm.

Conclusion

We offer our comments in recognition of the important contributions Ranulph Glanville made to research in architecture. He already recognized the difficulties such research faces as part of the work for his PhD. Later he was a forceful advocate of dealing with this difficulty in a very serious way. He proposed linking it to second-order cybernetics as an approach that is on the research side of Popper's demarcation problem. The value of his contributions is not diminished in any way by the fact that others have been trying to resolve the same difficulty; it is highlighted by the long-term efforts we also had to spend to go beyond approaches that aim to support professional developments, rather than serve as research.

In this context we wish to point also to the nascent field of artistic research (i.e., efforts to develop the notion of knowledge or competence being derived from artistic practice per se)—a field Ranulph touches upon occasionally. We see a similar difficulty, that is, the need to make it possible for others to accept the results of efforts to cooperate—those constrained by the interactions and conversations that take place given differences in individual preferences as well as resources. Where suitably organized such efforts will help achieve what research is intended to do: to help identify justifiably what resources can be recognized and used to develop artists' intuitions and intentions.

Our description of efforts similar to Ranulph's may help raise awareness of the difficulties encountered. The approach we choose is to identify constraints on what those engaged in interaction may contribute, in terms of other people's contributions. Some constraints seem to lock people into particular ways of talking (e.g., consumerism, cults of various sorts, including those self-created by artists). Others allow for freedom and variation (such as genuinely creative communities). Research may contribute to both approaches. We obviously prefer the second.[13] It links to the future—a way of thinking Ranulph made his own. In turn his work links to what we, and surely others after us, will continue, in the hope of opening up freedoms and opportunities for architects, artists, researchers and others.

References

Agar, M. (2013). *The lively science. Remodeling human social research.* Minneapolis, MN: Mill City Press.

Arrow, K. J. (1950). A Difficulty in the Concept of Social Welfare. *The Journal of Political Economy, 58*(4), 328-346.

Axelrod, R. (1984). *The evolution of cooperation.* New York: Basic Books.

Bertalanffy, L. von (1968). *General system theory: Foundations, development, applications.* New York: George Braziller.

Descartes, R. (2005). *Discourse on the method of rightly conducting the reason and seeking for truth in the sciences* (E. S. Haldane, Trans.). Stilwell, KS: Digireads. (original date: 1637)

Flood, R. L., & Jackson, M.C. (1991). *Creative problem solving: Total systems Intervention.* Chichester, UK: Wiley.

Foerster, H. von (1995). *The cybernetics of cybernetics* (2nd edition). Minneapolis, MN: Future Systems.

---

13. Recent PhD studies in the area of architecture demonstrate a similar preference, for example those of Janssens (2012) and Hendrickx (2012).

Galilei, G. (1989). Sidereus Nuncius or the Sidereal Messenger. Chicago: University of Chicago Press. (original date: 1610)

Glasersfeld, E. von (1984). An introduction to radical constructivism. In P. Watzlawick (Ed.). *The invented reality* (pp. 17-40). New York: Norton.

Glanville, R. (1975). *A cybernetic development of theories of epistemology and observation, with reference to space and time, as seen in architecture.* Unpublished PhD thesis, Brunel University. London.

Glanville R. (1988). *Objekte.* Berlin: Merve Verlag.

Glanville, R. (2002). Second-order cybernetics. In *Encyclopaedia of Life Support systems.* Oxford: EOLSS Publishers.

Glanville, R. (2004). The purpose of second-order cybernetics. *Kybernetes, 33*(9/10), 1376-1386.

Glanville, R. (2005). A (cybernetic) musing: Certain propositions concerning propositions. *Cybernetics and Human Knowing, 12*(3), 87-95.

Heidegger, M. (1962/2001). *Being and time* (J. Macquarrie & E. Robinson, Trans.). Oxford, UK: Blackwell.

Hendrickx, A. (2012). *Substantiating displacement.* PhD thesis. Royal Melbourne Institute of Technology, Melbourne.

Hughes, R. (2014). Exposition. In H. Borgdorff & M. Schwab (Eds.), *Exposition in artistic research.* Leiden, Netherlands: University of Leiden Press.

Hughes, R. (2012). Belgium Conversations 2011-2012. Reflections 16.

Hughes R., Anstey T., Grillner K. (Eds.) (2007). Architecture and authorship. London: Black Dog Publishing.

Hughes, R., & Monk, J. (Eds.) (1998). *The book of models: Essays on ceremonies, metaphor and performance.* Milton Keynes, UK: Open University.

Janssens, N. (2012). *Utopia-driven projective research.* PhD thesis. Gothenburg: Chalmers University.

Kuhn, T. S. (1962). *The structure of scientific revolutions.* Chicago: University of Chicago Press.

Lakatos, I. (1978). *The methodology of scientific research programmes.* Cambridge, UK: Cambridge University Press.

Latour, B. (1999). Pandora's hope. Cambridge, MA: Harvard University Press

Lewin, K. (1948). *Resolving social conflicts; selected papers on group dynamics* (G.W. Lewin, Ed.). New York: Harper & Row.

Mangan, J. A. (1986). *The games ethic and imperialism: Aspects of the diffusion of an ideal.* Harmondsworth, UK: Viking.

Marx, K. (2013). *Das Kapital.* Beijing: Intercultural Press.

Maskin, E. S. (2008). Mechanism design: How to implement social goals. Nobel Prize lecture, 2007. Reprinted in *American Economic Review, 98*(3), 567-576.

Neumann, J. von, & Morgenstern, O. (2000). *Theory of games and economic behaviour.* Princeton, NJ: Princeton University Press. (Originally published in 1944)

Pask, G. (2011). *The cybernetics of self-organisation, learning and evolution.* Vienna: echoraum.

Pask, G. (1975). *The cybernetics of human learning and performance.* New York: Hutchinson.

Popper, K. (2002). *The logic of scientific discovery* (2nd ed.). London: Routledge. (Originally published 1959)

Research Grants programme (2012). *Good practice guide: User involvement.* See: www.biglotteryfund.org.

Reason, P. & Bradbury-Huang, H. (Eds.). (2013). The SAGE handbook of action research: Participative inquiry and practice. London: Sage publications.

Rosen, R. (1991). *Life itself: A comprehensive inquiry into the nature, origin and fabrication of life.* New York: Columbia University Press.

Rosenhead, J. (2005). Past, present and future of problem structuring methods. *Journal of the Operational Research Society, 57*(7), 759-765.

Roshchupkina, M. (2015). *Research methods in fashion clothing retailing.* Unpublished doctoral dissertation, Lincoln University, UK.

Stachowiak, H. (1969). *Denken und Erkennen im Kybernetischen Modell.* Vienna: Springer.

Suppes, P., & Zinnes, J. L. (1963). Basic measurement theory. In R. D. Luce, R. R. Bush, & E. Galanter (Eds.), *Handbook of mathematical psychology.* New York: Wiley.

Vahl, M. (1994). *Improving mental health services in Calderdale.* Hull, UK: Centre for Systems Studies, University of Hull.

Wittgenstein, L. (1986). *Philosophical investigations* (3rd ed.; G. Anscombe, Trans.). Oxford, UK: Basil Blackwell.

Zeeuw, G. de (2006). A forgotten message? Von Bertalanffy's puzzle. *Kybernetes, 35*(3/4), 433-441.

Zeeuw, G. de (2007). The heroes and the helpers. In R. Glanville (Ed.), *Gordon Pask's legacy* (pp. 143-151). Vienna: Remaprint.

Zeeuw, G. de (2010). Research to support social interventions. *Journal of Social Intervention: Theory and Practice, 19*(2), 4-24.\

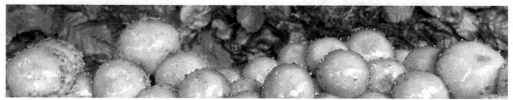

Bunnell, P. (2014). *Fungal Jewels.* Un-retouched photograph.

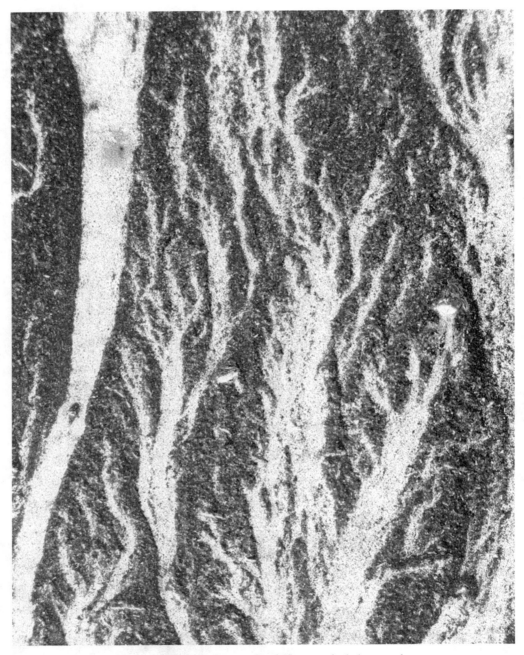

Bunnell, P. (2004). *Ferning Sand*. Un-retouched photograph.

Cybernetics and Human Knowing. Vol. 22 (2015), nos. 2-3, pp. 99-105

Conversation, Design and Ethics:
The Cybernetics of Ranulph Glanville

Ben Sweeting[1]

One of the major themes of Ranulph Glanville's work has been the intimate connection between cybernetics and design, the two principle disciplines that he has worked in and contributed to. In this paper I review the significance of the analogy that he proposes between the two and its connection to his concerns with, firstly, the cybernetic practice of cybernetics and, secondly, the relation between cybernetics and ethics. I propose that by putting the cybernetics-design analogy together with the idea that in cybernetics epistemological and ethical questions coincide, we can understand design as not just a form of cybernetic practice but also one in which ethical questions are implicit.
Keywords: cybernetics, design, ethics, conversation theory

Introduction

One of the major themes of Ranulph Glanville's research, and especially so in the period I have known him (from 2005 onwards), has been the intimate relation between cybernetics and design, the two principle disciplines that he has worked in and contributed to. The close analogy he developed between the two (Glanville, 2006a, 2006b, 2007c, 2009b), in terms of their conversational circularity, is notable not just for bringing the two disciplines to bear upon each other but also for unifying these two major strands of his own research.

In this paper I focus on this analogy and its significance. As well as noting some of the ways in which each discipline has the potential to inform the other, I emphasise connections with two of Ranulph's other major concerns within cybernetics: (1) with answering Margaret Mead's (1968) challenge that we practice cybernetics in accordance with its ideas; and (2) with the relation between cybernetics and ethics, on which Ranulph has written in this journal as well as elsewhere (Glanville, 2005, 2004/2009), and with which I was concerned in my PhD research as Ranulph's student (Sweeting, 2014).

Cybernetics and Design

I worked as research assistant to Ranulph when he guest-edited the *Cybernetics and Design* special double issue of *Kybernetes* (Glanville, 2007a) which set out to build bridges between the two disciplines.[2] In this intention, the issue followed on from many earlier examples of crossovers and influence. In architecture, my own field, the

---

1. Ben Sweeting, School of Art, Design and Media, University of Brighton. Email: R.B.Sweeting@brighton.ac.uk
2. In this I was funded by the Architectural Research Fund of the Bartlett, University College London.

most notable examples centre on the work of Gordon Pask, Ranulph's mentor.[3] During the 1960s and 1970s Pask collaborated with the architect Cedric Price and theatre director Joan Littlewood on the Fun Palace project and with Nicholas Negroponte's Architecture Machine Group at MIT, while his Musicolour installation was a key influence on John and Julia Frazer's contribution to Price's Generator project.[4] Since that period the relation between the two fields had become obscured and, while the increasing possibilities of technology have led to a reawakening of interest in cybernetics amongst designers, this has often been, as Ranulph notes, in rather "ancient" (Glanville, 2007c, p. 1177) cybernetics rather than an up to date understanding of the field. In organising this collection, and especially in his own contribution, Ranulph attempted to reconnect each field to the other, introducing cyberneticians to design and designers to cybernetics.[5]

If one of the motivations for the journal issue as a whole was with (re)establishing the potential relevance of each field to the other, in the second half of his paper Ranulph proposed seeing them not just as being related but as closely paralleling each other, to the extent that "cybernetics is the theory of design and design is the action of cybernetics" (Glanville, 2007c, p. 1178). This claim rests on a close analogy between the sort of conversational circularity with which cybernetics is concerned, as explored especially in Pask's (1976) conversation theory, and the distinctive way in which designers work, as is particularly evident in characteristic methods such as sketching as well as more generally in modelling and drawing. Indeed, the way designers work is often understood as a conversation that they hold with themselves and others, such as, for instance, in the account given by Donald Schön, who characterises design as a "reflective conversation with the situation" (Schön, 1983/1991, p. 76).[6]

The activities of conversation and sketching both share a circular form. In sketching, this circularity is created by our shifting perspective between looking and drawing in a way that parallels the turning around between listening and speaking in a conversation.[7] This enables evaluations of previous actions to influence present ones in a classic example of cybernetic feedback. The significance of this structure, however, goes beyond a cycle of iterations in pursuit of some goal. Just as a conversation, because of its interactive structure, tends to lead somewhere we could not have predicted in advance, so too the conversational interactivity of designers'

---

3. Pask held a post as a consultant at the Architectural Association and it was here that he introduced Ranulph to cybernetics.
4. As John Frazer himself notes (Furtado Cardoso Lopes, 2008, p. 58).
5. It strikes me now, in writing this, that Ranulph was deeply concerned with making introductions, both in this sense and between people.
6. See also the accounts of Cross (2007), Gedenryd (1998) and Goldschmidt (1991). Ranulph also notes a number of other parallels, such as a shared attitude towards error as neither good nor bad but endemic and a mutual concern with constructing the new, but these can each be traced back to the conversational analogy.
7. The etymology of conversation reflects this, to converse being "to turn about with.".

drawing and modelling is one way in which they create novelty[8] and is what allows them to work in the complex situations which they commonly encounter.[9]

This tendency towards the new in a conversation follows from the way that meanings are not transferred between participants but, rather, participants construct their own understanding of the understanding of others, with the process taking the recursive form of "what I think of what you think I think, etc." (Glanville, 1993, p. 217). For instance, if, in a simple two person conversation, I begin by presenting some idea, the other participant does not simply have this transferred to them but builds their own understanding of what it is that I mean. They then present what they have understood back to me and, again, I construct my own understanding of their presentation. I can then compare this understanding (what I understand of what they understood) to what I originally meant to communicate (see the diagram given in Glanville, 2009b, p. 432).

While we can repeat this process in an attempt to align these separately constructed understandings, it is not just—and often not even[10]—a way to reach agreement about existing ideas but, and more significantly in terms of the parallel with design, a way to generate new ones. Given that the difference between what participants understand is maintained throughout, we construct new understandings at every turn. We learn from the ideas that others present to us and from their comments on and criticisms of our own thoughts. We also often learn through misunderstanding, where we see a worthwhile idea in what someone says that was not intended. Perhaps most simply, we learn what is implied by our own ideas by seeing how they are interpreted and understood by others. This also occurs in conversations we hold with ourselves, as for instance in sketching.[11] When sketching, the designer simultaneously plays the roles of speaker (drawing) and listener (looking), switching roles between the two. Looking at what they have drawn they see some new possibility not previously intended and which can be developed further (Glanville, 2007c, p. 1189). In this sense designers' drawing is an integral part of their thinking, not a representation of ideas constructed previously.

While in one sense the feedback process of sketching allows designers to pursue some idea, as with conversation, this idea is not fixed at the outset but is developed through the process.[12] As Denys Lasdun put it, in remarks which are often quoted (not least by Ranulph), the architect's "job is to give the client, on time and on cost, *not*

---

8. The importance of novelty to designers follows from their concern with transforming existing situations into new ones. Ranulph also associates novelty with delight, one of the key characteristics which architects try to achieve in their designs, dating back to Vitruvius' *Ten Books on Architecture* (I.iii.2, trans. 1624).

9. On design as an approach to complexity, see Glanville, 2007b; 2007c, pp. 1195-1196; 2011a).

10. While we can try to reach agreement we will often abandon the attempt either through frustration or, alternatively, through the agreement to disagree (Pask, 1988, p. 85).

11. In Pask's conversation theory, the psychological-individuals, the participants in a conversation, are not in one-to-one relationship with the mechanical-individuals in which they are embodied. One may therefore have a conversation with oneself (as in sketching) taking different points of view in turn, while a group or organisation may act as one participant.

12. That is, the conversational circularity of cybernetics is not optimisation, which, as Negroponte has noted, is something "extremely antagonistic to the nature of architecture" (Negroponte, 1975, p. 189).

what he wants, but what he never dreamed he wanted and when he gets it, he recognizes it as something he wanted all the time" (Lasdun, 1965, p. 185). While this is sometimes read as a paternalistic claim to genius or expertise, it is better thought of as indicating the way that, in trying to achieve some idea, we revise not just the attempt to fulfill it but also the idea itself, having learnt more about it and the situation just as a conversation changes course through our participation in it.[13] Indeed, Lasdun's remark can be extended to apply as much to designers as to their clients, with new possibilities created as part of the process which designers could not have anticipated in advance.

From Cybernetics to Design and Design to Cybernetics

The point that Ranulph makes with this analogy concerns what is special and distinctive about design. As such, bringing cybernetics and design together is not about changing what designers do but, rather, about recognising the value in what is already being done and supporting it. In this it is different to many of the encounters that design has had with theory, where the task of theory is taken as correcting or reforming design in new ways. This is especially the case with architecture, which has a tendency to import theories external to itself in an attempt at reinvention or justification, often with the consequence of distorting the questions at hand.[14] In contrast, rather than cybernetic theory being something to be applied to design, the analogy with conversation positions cybernetics and design as coinciding with each other, a point that is reinforced by the designerly nature of much cybernetic research, such as that of Pask, where ideas are explored performatively through the construction of experimental devices (a quality which has been emphasised by Pickering, 2010). This reinforces design's disciplinary foundations, articulating its implicit epistemology (Glanville, 2006a, 2006b) and helping us see the sometimes contested relation between design and research in designerly terms (Glanville, 1981, 1999).

The nature of this analogy means that, as well as bringing cybernetics to bear on design, design can also contribute to cybernetics, thus informing theory from practice as well as practice from theory (Glanville, 2014). While there are many ways in which ideas from design practice are relevant to cybernetic questions, perhaps the most significant of these is in answering the challenge Mead (1968) set the American Society for Cybernetics (ASC) at its inaugural conference, that of applying cybernetic ideas to cybernetic practice. This was of increasing importance to Ranulph in recent years, in his role as president of the ASC, where he proposed alternative conversational conference formats as well as running a competition in response to Mead's question, won by Mick Ashby (Glanville, 2011b, 2012, pp. 197-210). While

---

13. In the same article, Lasdun also notes that the "worst work our office has ever produced" is the "competition work where there is a programme which is half-baked and there is no exchange of ideas" (Lansdun, 1965, p. 195); i.e., where there is no client with which to converse (see also Cross, 2007, p. 52; Glanville, 2009b, p. 427).
14. This is most notably the case with the misguided attempt of the Design Methods Movement to scientise the design process; see Gedenryd (1998). See also Glanville (2004).

the most obvious legacy of Mead's challenge was the epistemological concerns of second-order cybernetics, as developed by Heinz von Foerster and others, in Ranulph's interpretation this question was more about the value of cybernetic performance, similar to that he sees in the way designers work. If we understand cybernetics and design as the theory and practice of each other, then the cybernetic practice of cybernetics must look very like design, as indeed do many of the examples that Pickering (2010) discusses. The significance of this parallel is heightened by the way that design is at home in often complex, ill-defined and ethically charged circumstances, demonstrating how we can act cybernetically in situations where such a suggestion might otherwise seem overly optimistic.

Cybernetics, Design and Ethics

Towards the end of his contribution to the special issue, Ranulph connects his cybernetics-design analogy to his earlier reflections on the relationship between cybernetics and ethics (Glanville, 2007c, p. 1197-1198). In his (Glanville, 2004/2009) article "Desirable Ethics," Ranulph argues that the central mechanisms of cybernetics—such as conversation, the black box (a long-standing concern of Ranulph's cybernetics; see Glanville, 1982, 2007/2009, 2009a), and others—depend on qualities that are commonly taken to be ethically desirable: generosity, honesty, learning, mutuality, open-mindedness, respect, responsibility, selflessness, sharing and trusting. The structure of this argument reflects that of von Foerster (1974/2003), who understood ethical considerations as coinciding with cybernetics. This is what underlies Ranulph's deep concern with both Mead's challenge and design practice: to act cybernetically, as Mead proposes and as designers do, is to act out these ethical qualities.

Ranulph's careful framing of this argument is worth noting. He avoids going beyond a descriptive account of ethics, referring to values that are seen as desirable rather than to what is ethical. In so doing he avoids the way that, as von Foerster (1992, p. 12) points out, speaking explicitly about ethics can lead to moralisation. What von Foerster proposes in response to this is that we keep ethical considerations implicit in our acting, avoiding articulating them explicitly. In particular he emphasises two points, both of which build on the inclusion of the observer in second-order cybernetics: taking complete personal responsibility and participating in dialogue with others (von Foerster, pp. 13-18). While von Foerster's account associates dialogue directly with language, we can also understand it in terms of design, as per Ranulph's analogy between design and conversation. In this way, putting the relationships between cybernetics and design and cybernetics and ethics together, it is possible to see design *via cybernetics* as an activity in which ethical considerations are implicit and, therefore, as a way of responding to the challenges which von Foerster raises (see Sweeting, 2014).

Conversation is, in words which von Foerster (1991) credited to Victor Frankl, a thinking "through the eyes of the other." Similarly, design, because of its conversational structure, involves the consideration of others.[15] While this is partly in its explicit attempts at participation such as, for instance, consultation with stakeholders, it is also implicit in the conversations that designers hold with themselves. When drawing, architects "walk through" their plans in order to imagine how what they have drawn would be experienced, putting themselves in the place of absent others, many of whom (passers-by, future users) they will not be able to meet (Kenniff & Sweeting, 2014; Sweeting, 2014, pp. 106-110). Whereas ethical considerations are often thought of as external to or even as in conflict with design, this consideration of others is indispensible to what designers do. This is not to say that the way designers act is necessarily ethically good but that it is, at least potentially, ethical in the sense of involving ethical considerations within it (i.e., in the sense that an ethical question is a question concerned with ethics rather than an ethically good question). In this way, design provides us with an example not just of cybernetic practice but also of a way of acting, applicable in complex circumstances, where ethical considerations are implicit.

Acknowledgements

This paper contains ideas developed as part of my PhD research which was funded by an AHRC Doctoral Scholarship and supervised by Neil Spiller and Ranulph Glanville, to both of whom I owe enormous thanks.

References

Cross, N. (2007). *Designerly ways of knowing*. Basel, Switzerland: Birkhäuser.
Furtado Cardoso Lopes, G. M. (2008). Cedric Price's Generator and the Frazers' systems research. *Technoetic Arts, 6*(1), 55-72.
Gedenryd, H. (1998). How designers work: Making sense of authentic cognitive activities. *Lund University Cognitive Studies, 75*. Retrieved November 6, 2015 from http://lup.lub.lu.se/record/18828/file/1484253.pdf
Glanville, R. (1981). Why design research? In R. Jacues & J. Powell (Eds.), *Design, science, method: Proceedings of the 1980 Design Research Society conference* (pp. 86-94). Guildford, UK: Westbury House.
Glanville, R. (1982). Inside every white box there are two black boxes trying to get out. *Behavioral Science, 27*(1), 1-11.
Glanville, R. (1993). Pask: A slight primer. *Systems Research, 10*(3), 213-218.
Glanville, R. (1999). Researching design and designing research. *Design Issues, 15*(2), 80-91.
Glanville, R. (2004). Appropriate theory. In J. Redmond, D. Durling & A. de Bono (Eds.), *Futureground: Proceedings of the Design Research Society international conference 2004*. Melbourne: Monash University.
Glanville, R. (2005). Cybernetics. In C. Mitcham (Ed.), *Encyclopedia of science, technology, and ethics*. Woodbridge, CT: Macmillan.
Glanville, R. (2006a). Construction and design. *Constructivist Foundations, 1*(3), 103-110. Retrieved November 6, 2015 from http://www.univie.ac.at/constructivism/journal/1/3/103.glanville
Glanville, R. (2006b). Design and mentation: Piaget's constant objects. *The Radical Designist, 0*. Retrieved November 6, 2015 from http://www.iade.pt/designist/pdfs/000_05.pdf
Glanville, R. (Ed.). (2007a). Cybernetics and design [Special double issue]. *Kybernetes, 36*(9/10).
Glanville, R. (2007b). Designing complexity. *Performance Improvement Quarterly, 20*(2), 75-96.
Glanville, R. (2007c). Try again. Fail again. Fail better: The cybernetics in design and the design in cybernetics. *Kybernetes, 36*(9/10), 1173-1206.

---

15. As well as our relationship to others, which I focus on here, this argument can be extended to personal responsibility (design having no right answers, we are ultimately responsible for what we decide) and the pursuit of goods (goals) internal to action (which is enabled by design's circularity).

Glanville, R. (2009). A (cybernetic) musing: Ashby and the black box. In *The black box, volume III: 39 steps* (pp. 365-373). Vienna: edition echoraum. (Reprinted from: *Cybernetics and Human Knowing, 14*(2/3), 189-196, 2007).

Glanville, R. (2009a). A (cybernetic) musing: Black boxes. In *The black box, volume III: 39 steps* (pp. 405-421). Vienna: edition echoraum. (Reprinted from: *Cybernetics and Human Knowing, 16*(1/2), 153-167, 2009).

Glanville, R. (2009b). A (cybernetic) musing: Design and cybernetics. In *The black box, volume III: 39 steps* (pp. 423-425). Vienna: edition echoraum. (Reprinted from: *Cybernetics and Human Knowing, 16*(3/4), 175-186, 2009).

Glanville, R. (2009). A (cybernetic) musing: Desirable ethics. In *The black box, volume III: 39 steps* (pp. 293-303). Vienna: edition echoraum. (Reprinted from: Cybernetics and Human Knowing, 11(2), 77-88, 2004).

Glanville, R. (2011a). A (cybernetic) musing: Wicked problems. *Cybernetics and Human Knowing, 19*(1-2), 163-173.

Glanville, R. (2011b). Introduction: a conference doing the cybernetics of cybernetics. *Kybernetes, 40*(7/8), 952-963.

Glanville, R. (Ed.). (2012). *Trojan Horses: A rattle bag from the 'Cybernetics: Art, design, mathematic – A meta-disciplinary conversation' post-conference workshop*. Vienna: edition echoraum.

Glanville, R. (2014). Acting to understand and understanding to act. *Kybernetes, 43*(9/10), 1293-1300.

Goldschmidt, G. (1991). The dialectics of sketching. *Creativity Research Journal, 4*(2), 123-143.

Kenniff, T.-B., & Sweeting, B. (2014). There is no alibi in designing: Responsibility and dialogue in the design process. *Opticon1826* (16). (doi: 10.5334/opt.bj)

Lasdun, D. (1965). An architect's approach to architecture. *RIBA Journal, 72*(4), 184-195.

Mead, M. (1968). The cybernetics of cybernetics. In H. von Foerster, J. D. White, L. J. Peterson & J. K. Russell (Eds.), *Purposive Systems* (pp. 1-11). New York: Spartan Books.

Negroponte, N. (1975). *Soft architecture machines*. Cambridge, MA: The MIT Press.

Pask, G. (1976). *Conversation theory: Applications in education and epistemology*. Amsterdam: Elsevier. Retrieved September 30, 2015 from http://pangaro.com/pask/ConversationTheory.zip

Pask, G. (1988). Learning strategies, teaching strategies, and conceptual or learning style. In R. R. Schmeck (Ed.), *Learning strategies and learning styles* (pp. 83-100). New York: Plenum Press.

Pickering, A. (2010). *The cybernetic brain: Sketches of another future*. Chicago, IL: University of Chicago Press.

Schön, D. A. (1991). *The reflective practitioner: How professionals think in action*. Farnham, UK: Arena. (Original work published: 1983).

Sweeting, B. (2014). *Architecture and undecidability: Explorations in there being no right answer—Some intersections between epistemology, ethics and designing architecture, understood in terms of second-order cybernetics and radical constructivism*. Unpublished PhD Thesis, University College London. Retrieved Sept 30, 2015 from https://iris.ucl.ac.uk/iris/publication/972511/1

von Foerster, H. (1991). Through the eyes of the other. In F. Steier (Ed.), *Research and reflexivity* (pp. 63-75). London: Sage.

von Foerster, H. (1992). Ethics and second-order cybernetics. *Cybernetics and Human Knowing, 1*(1), 9-19.

von Foerster, H. (2003). Cybernetics of epistemology. In *Understanding understanding: Essays on cybernetics and cognition* (pp. 229-246). New York: Springer-Verlag. (Original work published in 1974).

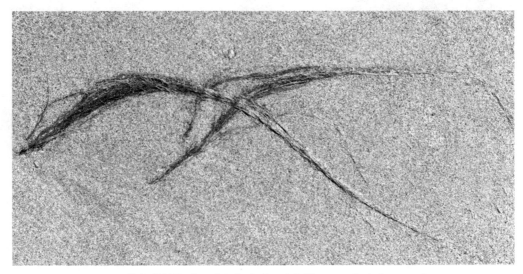

Bunnell, P. (2015). *Ranulph Said: Simplify.* Un-retouched photograph.

Bunnell, P. (2007). *Cold Heat*. Un-retouched photograph.

Cybernetics and Human Knowing. Vol. 22 (2015), nos. 2-3, pp. 107-114

The Big Picture:
Connecting Design, Second-order Cybernetics and Radical Constructivism

Christiane M. Herr[1]

This paper discusses the close relationships between design, second order cybernetics and radical constructivism that Ranulph Glanville has identified in his writings over the past decade. In linking these three fields, Glanville has established an overarching picture that shows how action, ethics and epistemology are related in a mutually complementing manner. While Glanville does not explicitly link all three fields in one dedicated paper, he elucidates one of these relationships each in three of his writings. In "Radical Constructivism = Second-order Cybernetics" (2012) Glanville asserts that second-order cybernetics and radical constructivism are "opposite sides of the same coin." Glanville lists seven core concepts of radical constructivism as stated by Ernst von Glasersfeld, and relates them to second-order cybernetic concepts. "Construction and Design" (2006) shows how design is a necessarily constructivist activity—both in terms of the design of concepts and the design of objects and processes. In "Try Again. Fail Again. Fail Better: The Cybernetics in Design and the Design in Cybernetics" (2007), Glanville presents second-order cybernetics as a theory for design, and characterizes design as cybernetics in practice. Drawing primarily upon these three papers, I construct a condensed version of Glanville's *big picture*. The value of the connections made lies in showing the role of each field in relation to the others, which both informs and affects each of the three fields thus connected.
Key words: design; second-order cybernetics; radical constructivism; Ranulph Glanville

Ranulph Glanville: Constructing the Conversational In-between

In various of his writings over the past decade, Ranulph Glanville has pointed out close relationships between design, second-order cybernetics and radical constructivism. In linking these three fields, Ranulph has transcended disciplinary divides and established an overarching picture that shows how action, ethics and epistemology are related in a mutually complementing manner. Ranulph's way of linking the three subjects is one of showing analogies across different subjects. Through carefully constructing such analogies, Ranulph creates a space that is less comparative than it is conversational: an in-between of different subjects that allows for mutually complementing productive exchanges and novel ideas. In this paper, I discuss these relationships based on three key papers: "Construction and Design" (Glanville, 2006), "Try Again. Fail Again. Fail Better: The Cybernetics in Design and the Design in Cybernetics" (Glanville, 2007), and "Radical Constructivism = Second-order Cybernetics" (Glanville, 2012).

Reading these three papers—all of Ranulph's works, for that matter—one is struck by the simple and clear manner in which he conveys his thoughts. Always

---

1. Email: Christiane.Herr@xjtlu.edu.cn

careful to build up arguments in a slow and rigorous manner, Ranulph maintains clarity while simultaneously including footnotes to a myriad of references offering further developments of the thoughts presented, as in leading multiple conversations at once. In addition, Ranulph is always keen to apply the contents of his writings to their form, creating consistency between argument and format in a cybernetic manner. This move —"application of understandings to self" (Glanville, 2012, p. 39)—forms the first core point in Ranulph's list of characteristics of second-order cybernetics as given in "Radical Constructivism = Second-order Cybernetics." Ranulph's often unusual writing style is a result of him enacting his understandings: His papers show as much as they communicate their contents. As Brier (2008, p. 84) puts it: "Glanville does not display knowledge but rather elicits knowing, exemplifying an educational practice directly reflective of his psychological and epistemological premises." The source of this preference underlies second-order cybernetics, which grew out of Margaret Mead's (1968) call to apply cybernetic ways of thinking to the study of cybernetics. Accordingly, Ranulph's writings typically do not follow templates but are designed to be specific to match and complement their particular concepts and contexts (to speak in architectural terms).

I do not attempt here to repeat the carefully crafted arguments contained in each of the three papers in full. This paper aims to illustrate and bring together three in-betweens that were central to Ranulph's work, and to examine—in brief—how Ranulph achieves the connections he makes, drawing emphasis to the cyclical manner in which the three subjects are related, and thus to demonstrate a design-cybernetic epistemology in action.

Second-order Cybernetics and Design

In "Try Again. Fail Again. Fail Better: The Cybernetics in Design and the Design in Cybernetics" (Glanville, 2007) Ranulph sets out to examine the relationship between cybernetics and design. Having been educated as an architect at the Architectural Association, a prestigious private architecture school in London, and having met cybernetics in the person of Gordon Pask during his studies of architecture, Ranulph maintained a deep interest in design throughout his career, teaching architecture in various universities and producing a large body of written work on the nature of designing. For this paper, I concentrate on Ranulph's positioning of cybernetics and design as complementary to each other, as he illustrates it in careful detail in his 2007 paper.

As a foundation for establishing connections between design and cybernetics, Ranulph takes the paper as an opportunity to mutually introduce the subjects to each other by first characterizing both cybernetics as well as design before engaging the subjects with each other. In this manner, Ranulph creates a conversational setting that allows readers familiar with one subject to get to know the respective other, thus providing a fertile ground for conversational exchange. The introduction further serves two other purposes close to Ranulph's heart: Firstly, to validate design as a subject when contrasted with science; and secondly, to introduce readers to the characteristics of second-order cybernetics, as distinct from first-order cybernetics. In introducing design and cybernetics as subjects, along with their relationships, Ranulph also warns readers of his "radical disagreement with many other authors" (Glanville, 2007, p. 1174).

Ranulph's characterization of design emphasizes the core act of designing, the doing that distinguishes design, and the source of the creativity and novelty associated with designing (Glanville, 2007, p. 1174). Insisting on the distinctive nature of design, Ranulph maintains that design is a "worthwhile act in its own right," (Glanville, p. 1179) and not to be forced into the corset of science as suggested by many other authors. Ranulph posits designing as both a way of acting and a way of thinking that has much to offer to other areas. Drawing parallels between the histories of design and cybernetics, Ranulph shows both subjects undergoing important transformations during similar periods: Attempts to make design conform to scientific ways of operating failed and began to be reconsidered around the same time when cybernetics became less traditionally scientific due to its consideration of the observer's role within observed systems (Glanville, p. 1176). These transformations resulted in both design and cybernetics "accepting the inescapable presence" and active role of the observer (Glanville, p. 1177). Despite these parallels, designers interested in cybernetics today often remain unaware of more recent developments in cybernetics, typically referring to first order approaches when approaching the subject.

Describing the design process for cyberneticians, Ranulph characterizes designing as a cyclical "conversation held mostly (but not exclusively) with the self" (Glanville, 2007, p. 1178). This characterization draws out the cybernetic nature of designing: "for conversation is perhaps the epitome of second order cybernetic systems" (Glanville, p. 1179). The conversational nature of designing allows design processes to be open to the unexpected and new, the creative. Designers typically sketch out their initially vague ideas, and in looking at the sketch, see it in new ways, as "viewing is an exploratory and constructive act" (Glanville, p. 1179). Sketching as the central source of creativity in designing, can thus "be described and explained as and by means of a primary second order cybernetic system—the circle of conversation" (Glanville, p. 1179). Ranulph positions cybernetics similarly to design, as a way of thinking that engenders a responsibility for one's own actions.

Exploring the implications of these analogies between design and cybernetics in the main body of the paper, Ranulph develops in detail the centrality of circularity to both subjects, defining cybernetics as the study of circular systems and their consequences (Glanville, 2007, p. 1200). Design, in turn, is a form of Paskian conversation and as such a cybernetic system. As a corollary of addressing circularity, both design and cybernetics transcend science. In the case of design, Ranulph argues that design is an inductive process focusing on action, whereas science is deductive and descriptive in nature (Glanville, p. 1179). Designing thus cannot be scientific—but science can be considered a restricted form of designing (Glanville, 1999). In the case of cybernetics, the consideration of observers as part of observed systems introduces circular causalities that are not admitted in conventional science (Glanville, 2007, p. 1182). In this context, Ranulph draws attention to Wiener (1948), who, aware of the implications of admitting circular causality, was careful not to call the subject of his famous book *Cybernetics* a science.

While the essential points I draw out of Ranulph's detailed argument here must remain brief, I hope to show the convincing analogy established between design and second order cybernetics: "cybernetics is the theory of design and design is the action of cybernetics" (Glanville, 2007, p. 1178).

Design and Radical Constructivism

In "Construction and Design" (Glanville, 2006) Ranulph sets out to construct a similar analogy between design and constructivism, arguing that designing is a constructivist activity preceding constructivism by millennia. Again, Ranulph does not aim to establish similarity that could dissolve (productive) differences between the two subjects. Instead, he aims to show how the presence of each within the other can enrich both subjects (Glanville, p. 103). The starting point for Ranulph's investigation is the individual level of experience. If we assume that every individual lives in a stream of experience that is essentially personal and private, the question arises of how we can explore the experience of the other? Moreover, how can we conceptualize means of communication (Glanville, p. 103)? Both questions feature prominently in radical constructivist theory (see Glasersfeld, 1995) and are the core of much epistemological debate. To designers, however, Ranulph argues, the existence of a "mind independent reality" does not matter much, as they are too busy to make things for their presumed "real" world. Ranulph argues that to him, the question whether a "mind independent reality" exists is in principle undecidable (Glanville, p. 103), as we can neither assert nor deny its existence. Given this ambivalence, we are free to choose our position depending on the circumstances. Designers choose both, embracing the paradox, whereas Ranulph himself asserts his choice to maintain undecidability in this question in order to keep the possibility of making the choice spontaneously.

Once more, Ranulph introduces design in "Construction and Design," but this time to the audience of the journal *Constructivist Foundations*, which can be assumed to be well versed in radical constructivist theory. Accordingly, Ranulph focuses on the constructive nature of designing, and characterizes designing as an activity that proceeds to construct new realities—without questioning the nature or conditions of the reality in which designers construct such realities. The central activity of designing, Ranulph argues, is the act of giving form to the unformed (Glanville, 2006, p. 104), and thus necessarily creative. As in his 2007 paper, Ranulph describes in detail the process of sketching as an illustration of the central act of designing: Sketching implies to first draw roughly, incompletely, then reflecting on and seeing anew the outcomes, as an ongoing circular activity of re-consideration. In this process, the initially aimless, "seemingly purposeless, playful and dreamy activity that is at the heart of design" (Glanville, p. 105) eventually produces a plethora of new objects – objects in the form of things and processes, but also conceptual objects. Design processes, Ranulph thus argues, are the fundamental human activity leading to new thoughts.

To further explain the design of concepts, Ranulph invokes the cybernetic device of the Black Box. The Black Box is a thought experiment that allows observers to generate descriptions and explanations of observed changes, without making any claim as to what "actually" happens (Glanville, 2006, p. 107). Observers can place an imaginary Black Box wherever desired, obscuring that which descriptions and explanations are sought for. Each Black Box is presumed to have inputs, and outcomes resulting from the inputs in conjunction with what happens inside the Black Box. Observers can then create descriptions that account for observations made, and hope that descriptions and explanations obtained in this manner will continue to account for what is observed. Ranulph identifies this mechanism in conversational processes that lead to the construction of both stable physical as well as mental objects, similar to the constant, conserved objects described by Piaget (1955) as populating each individual's world of experience. "In this view, we design our world. We create our concepts from which we can conserve our objects. This is our world. It is designed. Thus, it is constructed" (Glanville, p. 107). Discussing the nature of knowledge thus obtained, Ranulph identifies two distinct types of knowledge: the conventional descriptive *knowledge of* and the action-oriented *knowledge for* that designers tend to pursue. The second type of knowledge is uniquely suited to our "world of action," and is in fact the type of knowledge that constructivists use—but it has so far been underestimated. Ranulph argues that it deserves (urgent) further study and development (Glanville, p. 108).

Radical Constructivism and Design

In "Radical Constructivism = Second-order Cybernetics" (Glanville, 2012), Ranulph focuses on bringing together radical constructivism and design, thereby completing the circle of relationships he develops between the three subjects. The title of the paper posits an equation, with which Ranulph intends to distinguish the two subjects as "opposite sides of the same coin" (Glanville, p. 27). As in the two papers reviewed above, this serves to bring together two subjects in the spirit of conversation rather than dissolving their distinctive natures. Ranulph observes this view to be common among second order cyberneticians, yet not as common for radical constructivists. Accordingly, Ranulph addresses a primarily radical constructivist audience to demonstrate how seven core concepts of radical constructivism—as stated by Ernst von Glasersfeld (2007)—can be found in second-order cybernetics. To do this, Ranulph illustrates Glasersfeld's core concepts by linking it to the work of various second order cyberneticians. In the following, I briefly summarize several of these points, along with Ranulph's second order cybernetic equivalents.

Radical constructivism sidesteps the concerns of conventional epistemologies and has been described as a post-epistemological approach (Noddings, 1990; Glasersfeld, 1990) in which knowledge is not taken to represent a world independent of and beyond experience. Radical constructivism thus introduces a modified concept of knowledge as pertaining to "the way in which we organize the world of our experience" (Glasersfeld 2007, p. 97). In a similar manner, Ranulph argues, second-order cybernetics is based on the acceptance of the observer in the system. In addition, Bateson's notion of the *explanatory principle* (Bateson, 1969) emphasizes that concepts such as gravity are invented by observers to explain and organize experiences. Maturana's work develops a similar position, arguing that experience and the nervous system are not connected but instead, coordinated (Maturana & Varela, 1998).

Glasersfeld (2007, p. 97) points out that radical constructivism cannot assert or deny the existence of a mind independent reality. Ranulph (Glanville, 2012, p. 31) invokes Heinz von Foerster's notion of the "undecidable question" (Foerster, 2003)—a question which is in principle undecidable requires the individual to make a decision, thereby also assuming responsibility for that decision. Already in his own PhD thesis, Ranulph states that "In order to know something exists we must be able to observe it" (Glanville, 1975, p. 15), and "if we cannot observe it, we cannot know it exists. We cannot necessarily affirm its non-existence, either" (Glanville, p. 233).

Radical constructivism is explicit about human knowledge being a human construction, with Piaget (1955) describing the constructive conceptual operations humans employ in the process of organizing their experiential worlds. Ranulph shows how analogous arguments were made by Foerster (1973) in "On Constructing a Reality" and Spencer-Brown's (1969) *Laws of Form*. Ranulph further invokes his own body of work concerned with design as the basic constructive act, and the cybernetic mechanism of the black box, as already discussed above.

Instead of the requirement of "truth" in the context of a mind-independent reality, radical constructivism requires of knowledge that it be viable. Viable knowledge fits into the world of the knower's experience and is maintained until it does not match experience. Ranulph (Glanville, 2012, p. 35) relates the radical constructivist notion of viability to Beer's notion of the viable system (Beer, 1972) as a system that can sustain its own autonomy, as well as to Varela, Maturana and Uribe (1974) who describe autopoietic systems in a similar manner.

Ranulph ends his paper with a call to his audience to complete the argument from a radical constructivist perspective by taking the nine core characteristics of second-order cybernetics Ranulph lists at the end of the paper and showing "how radical constructivism might fit under headings that describe second order cybernetics." (Glanville, 2012, p. 39) To my knowledge, this call has yet to be pursued further and may be the last part still missing in the big picture Ranulph creates in the three papers discussed here.

Summary and Open Questions

In this paper, I have shown how Ranulph Glanville has identified and illustrated interrelations of design (action), second-order cybernetics (ethics) and radical constructivism (epistemology) in his recent body of work. I have based my argument primarily on three of Ranulph's papers, which each develop one mutual connection between the subjects of design, second-order cybernetics and radical constructivism. The above three sections have constructed a circular set of analogies, exploring relationships between subjects first in pairs and finally ending where the analogy drawing began, to possibly begin anew. One part however still missing to complete the cycle of analogies is to show how core concepts of second-order cybernetics may be exemplified with radical constructivist concepts—as suggested at the end of "Radical Constructivism = Second-order Cybernetics" (Glanville, 2012).

In connecting the three subjects, Ranulph engages readers in a pattern-making exercise: We are invited to find how patterns we identify as characteristic of one subject can be also be thought of in terms of patterns taken from another subject's intellectual tradition, resulting in compelling—and often delightful—analogies. Reading Ranulph's carefully composed illustrations, we are as much convinced as we are invited to join in the pleasure of finding, of constructing our own new patterns. It is, finally, this conversational, somewhat open-ended way of writing that both employs and embodies Ranulph's appreciation of the conversational in-between in which new and surprising things can happen. This could be identified as one of the core characteristics of Ranulph Glanville's design cybernetics—if this list were to be composed in the future.

References

Bateson G. (1969). Metalogue: What is an instinct? In T. Seboek (Ed.), *Approaches to animal communication.* The Hague: Mouton. (Reprinted in: Bateson G. [2000]. *Steps to an ecology of mind* [2nd ed.; pp. 38-58]. Chicago: Chicago University Press.)

Beer, S. (1972). Brain of the firm. Harmondsworth, UK: Allen Lane.

Brier, S. (2008). Ranulph Glanville: The cybernetician of ignorance. *Cybernetics and Human Knowing, 15*(1), 81-90.

Foerster, H. v. (2003). Ethics and second-order cybernetics. In *Understanding understanding* (pp. 287-304). New York, NY: Springer.

Foerster, H. v. (1973). On constructing a reality. In F. Preiser (Ed.), *Environmental research* (Vol. 2, pp. 35-46). Stroudsberg, PA: Dowden, Hutchinson & Ross.

Glanville, R. (1999). Researching design and designing research. *Design Issues, 13*(2), 80-91.

Glanville, R. (2006). Construction and design. *Constructivist Foundations, 1*(3), 103-110.

Glanville, R. (2007). Try again. Fail again. Fail better: The cybernetics in design and the design in cybernetics. *Kybernetes, 36*(9/10), 1173-1206.

Glanville, R. (2012). Radical constructivism = Second-order cybernetics. *Cybernetics & Human Knowing, 19*(4), 27-42.

Glasersfeld, E. von. (1990). An exposition of constructivism: Why some like it radical. In R. B. Davis, C. A. Maher, & N. Noddings (Eds.), *Constructivist views on the teaching and learning of mathematics. Journal for Research in Mathematics Education Monograph #4* (pp. 19-29). Reston, VA: National Council of Teachers of Mathematics.

Glasersfeld, E. von. (1995). *Radical constructivism.* London: Falmer Press.

Glasersfeld, E. von. (2007). Aspects of constructivism. Vico, Berkeley, Piaget. In E. von Glasersfeld, M. Larochelle, E. Ackermann, & K. G. Tobin (Eds.), *Key works in radical constructivism* (pp. 91-99). Rotterdam, Netherlands: Sense.

Maturana, H. & Varela, F. (1998). *The tree of knowledge.* Boston: Shambala.

Mead, M. (1968). The cybernetics of cybernetics. In H. von Foerster, L. J. Peterson, & J. K. Russel (Eds.), *Purposive systems* (pp. 1-11). New York: Spartan Books.

Noddings N. (1990). Constructivism in mathematics education. In R. B. Davis, C. A. Maher, & N. Noddings (Eds.), *Constructivist views on the teaching and learning of mathematics. Journal for Research in Mathematics Education Monograph #4* (pp. 7-18). Reston, VA: National Council of Teachers of Mathematics.

Piaget, J. (1955). *The child's construction of reality.* New York: Basic Books.

Spencer-Brown, G. (1969). *Laws of form.* London: George Allen and Unwin.

Varela, F., Maturana, H. and R. Uribe (1974). Autopoiesis. *BioSystems, 5*(4), 187-196.

Wiener N. (1948). *Cybernetics, or communication and control in the animal and the machine.* Cambridge, MA: The MIT Press.

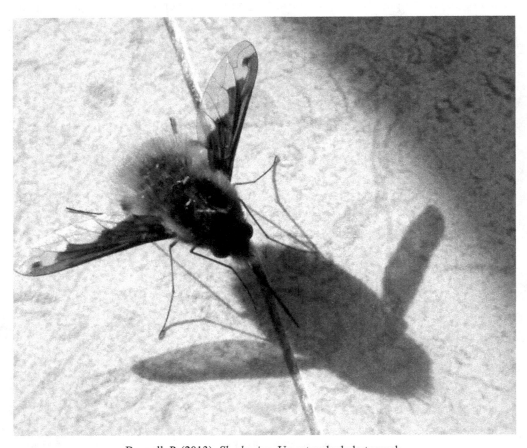

Bunnell, P. (2013). *Shadowing.* Un-retouched photograph.

Cybernetics and Human Knowing. Vol. 22 (2015), nos. 2-3, pp. 115-120

What I Learned from Ranulph Glanville

Larry Richards[1]

This paper is a personal statement about some of what I learned through my association and experiences with Ranulph Glanville. I focus on five aspects of this learning: the clarity with which he expressed complex ideas, his insistence on listening (as distinct from observing), his style of getting things done, his embodiment of conversation, and his special interest in design and the difficulty of language. There is much more that I could report; Ranulph made so many contributions to so many fields of study that it is only possible to scratch the surface. My personal top five will have to do for now.

Keywords: clarity of thought, listening, quiet determination, conversation, design and language

Introduction

Reflecting on how different people have influenced the path I have taken in my career and life has provided me with a continuing source of wonder. I would not be using the word wonder here if I had not heard Ranulph Glanville give a presentation on that topic at a conference in Vienna, Austria (Fall, 2003). I do not recall the presentation itself as particularly memorable, but Ranulph had a way of delivering a presentation that offered some idea or nuance that has then resonated in my mind ever since. This is a different sort of contribution than what others have offered me and one for which I will be forever grateful. I think it was Ranulph's tenacity and persistence that I am still trying to learn. He was never reluctant to talk about whatever was on his mind, no matter how controversial or unpopular, and he would address directly anyone who implied he should not be talking about a particular topic or in a particular way. He knew what he wanted to have happen, and nothing was going to stop him except his own change of mind.

I would not say I was close to Ranulph. The closest we got was when he visited Jane and me at our home in the spring of 2011. We were planning a conference in my home town for that summer, and he was insistent on seeing the hotel and other possible arrangements ahead of time. Again, he had an idea about what he wanted to happen and wanted to bring me in on it by visiting in person. I would like to think that he also wanted to get to know me a little better. It was during this visit that we had a chance to talk about his experiences with Alcoholics Anonymous, a subject I have since researched and to which I now refer often as an example of a non-hierarchical (even anarchical) organization. Below, I offer a few more examples of what I learned from Ranulph. My memories of him and our conversations continue to contribute to my wonder, influencing my path every day as I walk it.

---

1. laudrich@iue.edu

Clarity of Thought

I first met Ranulph when he showed up at a pre-conference tutorial on cybernetics in 1981 (American Society for Cybernetics, Washington, DC). It was not long after his arrival that he began contributing to the conversation. I did not know who he was until Stuart Umpleby, who was facilitating the tutorial, recognized him by name. I have been in these types of conversational events where interventions by someone who had just recently arrived, after the others had been there already a half a day, would not have been welcomed. However, Ranulph's comments demonstrated a clarity of thought that most in the room apparently took as going directly to the point (at least I did, and others did not complain). I still invoke Ranulph's way of commenting as I try to develop my own clarity of thought with respect to cybernetics.

I only later in the conference learned that Ranulph had been a student of Gordon Pask. I found this odd, as I would never have used clarity of expression as a way of describing Gordon's interventions in group conversations, although I would later learn that he was quite clear in his thinking. In fact, it took me three eight-week sessions (1987-88), working with Gordon during the afternoons and then talking with him over dinner almost every evening, to begin to understand what he had to offer. I recall the effort I had to exert to pay attention to what he was saying in order to sustain a conversation, a somewhat surprising (even paradoxical) exercise when interacting with someone who had coined the term *conversation theory*. I am sure that Ranulph would agree that Gordon was a special person in many ways, certainly unique by our experiences. In my case, I took on the task of trying to grasp the complexity of Gordon's dialogic and its contribution to the human predicament and then to find ways to express it that would not require the same level of effort and time that I had put into it. While I am still struggling with that task, it may perhaps be Gordon's style that has inspired many of his students to seek clarity in the expression of their thinking. However, I never experienced Ranulph compromising on the ideas or rendering them simplistic. It is that aspect of Ranulph's clarity that I seek to emulate.

Listening

During my first few encounters with Ranulph, it wasn't clear to me if he was really listening to what I or others were saying. He was and remained headstrong, and that characteristic may have overshadowed others for me during the early 1980's. Over time, however, I came to appreciate, not only that he was paying attention, but that he was listening as intently as anyone whom I had ever met. It was especially during the past 15 years, the last six of which he served as President of the American Society for Cybernetics (ASC), that his insistence on listening, for which he set an unmistakable example, inspired me to try to learn the skill by observing (and listening to) him.

At the American Society for Cybernetics (ASC) meeting in Troy, New York (Summer 2010), Ranulph engaged the entire conference in a listening exercise: After a practice session in which each participant created a humming sound, with variations, and then used it to respond to others next to them, Ranulph took us into a new grand concert hall at Rensselaer Polytechnic Institute. There, in the front of the hall, the entire (almost) conference performed a composition based on a few simple instructions requiring us to listen to each other. What I experienced was, over a period of about an hour—from the start of the practice session to the end of the performance, an enhanced ability to listen to and distinguish a wide range of individual sounds. Ranulph claimed he could hear every single individual (and name them), and I have no doubt that he could.

As ASC President, Ranulph asked if I would help organize the next conference (Summer 2011) in my home town of Richmond, Indiana. The theme would be "Listening." This opened up new possibilities of exploration on this topic, a gesture that brings a smile every time I think of it. The idea that listening might play a central role in an area of inquiry that had, for so many years, placed observing at its center, was intriguing, to say the least. The conference scheduled numerous performances (including one of Ranulph's compositions) that were oriented toward the activity of listening (including those performances that were movement-based). Observing with one's ears and listening with one's eyes (and nose, mouth and skin) became a real possibility; observing and listening became distinct activities in my cybernetics, each important and together revolutionary.

Quiet Determination

I did not learn the idea of quiet determination from Ranulph; I had been reading and studying about it as a style of leadership for years. What I learned from Ranulph was how it could be enacted. I had served a three-year term as ASC President (1986-88) and so appreciated the difficulty in getting such a disparate group of people—different disciplines and professions, different interests, different backgrounds, different preferences—together to advance the organization and the ideas of cybernetics. It seemed to demand a strong, charismatic leader, one who would be out in front advocating and pushing the envelope, irrespective of what anyone else wanted. While Ranulph was strong and charismatic, and definitely pushed the envelope as ASC President, he did not do so in a flamboyant and autocratic manner. He brought in many others, both for the organization of the conferences and for the operation of the Society. His determination was directed at what he knew would be in the interest of the ASC membership, current and future, and he worked many hours to ensure that key people knew what he was doing and why. I am sure it was like a full-time job for him, yet he never let on.

It would be inaccurate to describe Ranulph as quiet. When I use the term quiet determination, I am speaking of a determination that operates behind the scene, in the background. Ranulph's determination was always there, but it would give way to the need to build relationships and to allow processes of participation to work out. It would rear its head when someone would challenge a rule or design feature or guideline that was needed to realize an idea that many others had already vetted.

However, I noticed that when a rule or guideline was not followed, as in a conference design, he would often let it go. I took this to reflect a realization that something new might emerge if some flexibility were allowed. I am still trying to learn how he knew when to allow some variation and when not.

Conversation

To say that Ranulph was about conversation would be an understatement. He was the embodiment of conversation. That he could organize three ASC conferences around conversations (2010, 2011, 2013), with no scheduled paper presentations, and have people coming back for more, still leaves me breathless. Some of the best conversations I have had with respect to cybernetics occurred during these conferences. I am not sure that everyone at the conferences appreciated the importance of these events. After all, not all conversations are inspiring. What Ranulph was trying to do, in his subtle way, was to teach us conversation by having us do it. The conversations that worked were the ones where the participants gave attention to the subtle, as well as the not-so-subtle, moments of conflict, friction, disagreement, and/or asynchronicity in the interactions of which they were a part. These moments serve as opportunities to introduce a new twist or question or idea that will sustain the conversation as it moves toward agreement (including agreeing to disagree) and synchronicity.

Some conversations worked for the participants and some did not. What I learned was that the prospect of a participative-dialogic society is less of a pipe dream than I had feared it might be; that was encouraging. However, I also learned that such an idea requires tending to those details that, if not tended to, can derail or dismiss conversations before they even get started. For example, Ranulph was the first ASC President since 1974 to successfully enlist academic journals (*Cybernetics and Human Knowing* and *Kybernetes*) to publish papers written for the conference by the attendees. To emphasize the conversational theme of the conferences, the papers were not scheduled for formal presentation. Evenings were made available for authors who wanted to present their papers to do so; many were simply satisfied with the potential for publication and could relax that nagging expectation of the institutions that were funding their attendance. This then freed time for deep conversations to occur and for participants to focus on them, rather than on their papers. I found these conversational conferences to sustain the hope that the idea of conversation would not necessarily decay into mere communication, at least not in the near future. That was a great relief. Again, it was Ranulph's tenacity and passion for conversation that made this possible.

Design and Language

Ranulph was educated as a designer, and he taught design. He would often talk about how cybernetics and design went hand in hand. Some of his ideas about design played out in the design of cybernetics conferences, although they were also obvious in other aspects of his work. One idea that particularly resonated with me was that of design without purpose. I have struggled with the word purpose for many years. Ranulph and I did not necessarily agree on the use of certain words, like needs, desires, and intentions. Purpose may have also been in that category, yet I think we agreed that design could (and perhaps should) occur without purpose, at least in the standard use of the word. The standard use of the word purpose makes it a synonym with the word goal. Although there are also different uses of the word goal, I use it in what I regard as its everyday use—when I wish to speak of the formulation of a desire as an end point to be achieved. Consciously selecting a goal or a set of goals as a way to initiate the activity of design is limiting, to say the least. One of Ranulph's colleagues in design and cybernetics, Thomas Fisher, and I have recently written a paper on constraint-oriented design as an alternative to goal-oriented design, in which we propose that desires be formulated as constraints. Ranulph was even more open-ended with his idea of design without purpose, suggesting that design can start with nothing more than putting a pen to paper and seeing what happens. I am still struggling with this idea—I prefer to see more self-awareness of desires; however, maybe the desire to put pen to paper is sufficient.

Among what I learned from these exchanges about the concept of design, and all of its associated words, was the difficulty of sustaining conversations about language itself. I came to appreciate Ranulph's reluctance to hold a conference on the theme of language. Once a conversation turns to the variety of ways that words are used, he would lament, it becomes a mess of disagreements about how they should or should not be used, disagreements that do not go anywhere. Conversations get derailed before they begin. Better that people just use words to talk about whatever they want to talk about and then clarify their use of words when questions or conflicts arise, rather than try to get agreement on the use of words ahead of time. I would still like to think that this could apply to a conversation about language itself, but I do keep Ranulph's admonitions in the back of my mind and have often been thankful to have them there.

Concluding Thought

There is, of course, much more that I could report on Ranulph's contributions to my thinking and that of many others. He influenced people through his ideas and through his presence. I learned from both. For example, while Ranulph was a self-proclaimed atheist, he still talked about a relationship with the spiritual. In my case, I am not a disbeliever, as I do not accept the desirability of the concept of belief in the sense of acknowledging ultimate truth: For me, disbelief is just another form of belief. However, Ranulph's ideas about spiritual experience (from a cybernetic perspective) resonated with my particular form of agnosticism, and I have on occasion used them.

As another example, Ranulph was worldly; he travelled around the world, learning something new everywhere he went. When he visited Richmond, Indiana, the first time, it seemed that he learned in one day more about the city that I had learned in seven years. He also worked to internationalize the American Society for Cybernetics

(ASC). At a recent ASC conference in Bolton, England (2013), I counted more attendees from countries outside North America than from within. This was, I am convinced, an intentional strategy on Ranulph's part, reflecting his conviction that any study or advancement of cybernetics has to take a global perspective. This I now acknowledge, thanks to Ranulph.

A final example: Ranulph would, in conversation, tend toward an anarchist view of government, and organizational structures in general, and would refer to Alcoholics Anonymous as a model for how alternatives might be possible. Yet, I observed him as respectful of most authority figures in formal organizations when he interacted with them. One might argue that he recognized the importance of their support in the current society, and therefore the need to show respect; however, I think he also recognized that they were as much victims of the current systems as anyone else, and there was nothing to gain by disrespecting them as people. I also have some anarchist leanings—namely, I would like to see an experiment with a form of government that is different from any that currently exists in nation-states around the world: one with rules, but no rulers. I regret that I never got around to talking with Ranulph about current economic systems and the apparent need for strong and hierarchical governments to regulate those systems in order to protect the people from their competitive, and therefore predatory, tendencies. So, I would propose, perhaps as a tribute to Ranulph, that we engage the question: What economic systems can we imagine that would be compatible with a secular, anarchist, world-wide society? I am sure there is still much we can learn from Ranulph.

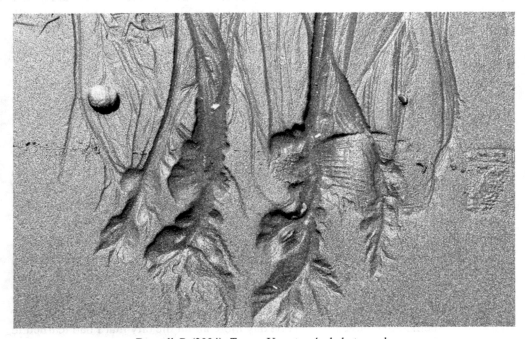

Bunnell, P. (2004). *Traces*. Un-retouched photograph.

Cybernetics and Human Knowing. Vol. 22 (2015), nos. 2-3, pp. 121-129

What I Learned from Ranulph:
A Grateful Tribute to Ranulph Glanville

Michael Lissack[1]

The first thing Ranulph Glanville taught me about cybernetics would prove in the end to be the most prophetic: namely that Winston Churchill was the first practicing cybernetician. I suspect that most if not all of the readers of this article are already shaking their heads. What could Lissack be talking about?

Ranulph told me this when describing a speech Churchill gave in the House of Lords chamber when the British Commons was discussing whether or not to rebuild the Palace of Westminster on October 28, 1943.

> On the night of May 10, 1941, with one of the last bombs of the last serious raid, our House of Commons was destroyed by the violence of the enemy, and we have now to consider whether we should build it up again, and how, and when.... We shape our buildings, and afterwards our buildings shape us. Having dwelt and served for more than forty years in the late Chamber, and having derived very great pleasure and advantage therefrom, I, naturally, should like to see it restored in all essentials to its old form, convenience and dignity. (Churchill & Churchill, 2003)

"We shape our buildings, and afterwards our buildings shape us." In that one sentence were a host of pearls of cybernetic wisdom—the role of context, the role of affordances, the importance of action, ideas shaping action, actions shaping ideas, circularity, and the role of the observer. And, Churchill said it five years before Weiner's first book. Thus, Churchill was the first practicing cybernetician (under modern usage—ignoring Ampere's 1834 definition).

Ranulph cautioned, however, that Churchill himself would have been opposed to the label. Which was Ranulph's second lesson to me: Many of those who are most successful in incorporating cybernetics into their world view and their affordances for action are highly resistant to the label cyberneticians and often to the very topic of cybernetics. These people live cybernetics. It is a part of who they are and of what they do. They do not study the subject. They do not write great tracts about critical ideas and alternative formulations. They do not spend time trying to sort out the vagaries of difference between systems science, complex systems, and cybernetics. They never heard of Science 1 and Science 2. And, they do not care. They live their lives. Cybernetically.

It was only fitting then that Ranulph's final ASC conference would be entitled "Living in Cyberentics." To the Ranulph I knew cybernetics was about acting, thinking, and then acting again. It was about life.

---

1. President, American Society for Cybernetics. Email: lissack@lissack.com

Paul Pangaro captured this idea in a video which I played at that 2014 conference:

> I believe that cybernetics is an exceptionally great way of characterizing how the world works. Where by world we mean the world that humans inhabit. Because we inhabit a mechanical physical world where things have to work. We inhabit a biological world which has to work (of course otherwise we would not survive). And a social world in which our conversations in which our conversations and interactions work as best they do. or, if they break down, cybernetics allows us to model the breakdown and to know how to improve things. I find it an incredibly powerful language... a frame for looking at the world. Once you see the world in a cybernetic way, through the cybernetic lens, all things are cybernetic. Because all systems become part of this set of languages of action and sensing and comparing and understanding and taking a meta-view. All intelligent systems have this property. of trying, acting, seeing the difference, changing, acting, seeing, sensing. This loop of acting, sensing, comparing is fundamental. (see https://www.youtube.com/watch?v=Cvorr7587dE)

Paul's view is approximately captured by the drawing in figure 1.

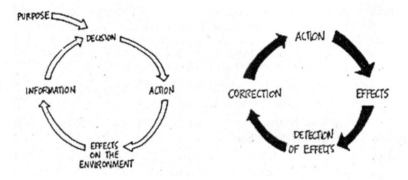

Figure 1: The Living Cybernetic Loop

When I asked Ranulph to explain his perspective to me, a five-year long seminar was begun (mostly in person, sometimes by phone, often by email). I can only summarize here what Ranulph taught me and I do so in Ranulph's own words:

> Second order Cybernetics presents a (new) paradigm in which the observer is circularly (and intimately) involved with/connected to the observed. The observer is no longer neutral and detached, and what is considered is not the observed (as in the classical paradigm), but the observing system. The aim of attaining traditional objectivity is either abandoned/passed over, or what objectivity is and how we might obtain (and value) it is reconsidered. In this sense, every observation is autobiographical. ... The principle of the Black Box is that, where we observe some change in a behavior, we construct and insert a Black Box allowing us to interpret the change as the result of the operation of an invisible mechanism, held within the Box, on what is now seen as input giving rise to output. The observer/scientist develops a description functioning as a mechanism/explanation (i.e., model) which accounts for the transformations of what are now input into output. The explanation is purely historical and the product of the interaction between the observer and his inventive, fictional insertion, the Black Box. ... What is vital, for the development of second

order Cybernetics, is that the Black Box is essentially and crucially a construct of the observer. When we use this concept, we bring the observer in to the process, rather than denying him. That the Black Box requires the observer's presence is acknowledged, and is circularly connected in. The observer watches and changes. What the observer learns he learns from interaction with the Black Box (which is his construct). ... When what is observed is observed by an observer, that observer is responsible for the observation, the sense he makes of it, and the actions he takes based on that sense. Von Foerster gives an Ethical Imperative: "Act always so as to increase the number of choices." (This is joined by an accompanying Aesthetical Imperative: "If you desire to see, learn how to act." The third is that we construct our realities. "Draw a Distinction!") ...

My major initial concern was to develop a set of concepts that might explain how, while we all observe and know differently, we behave as if we were observing the same thing. ... To use a metaphor: my work is the creation of games fields: others create the games to play in these fields and still others play them. Finally, some are spectators. The point of an account that admits others is not that it is right, but that it is general (and generous). Cybernetics is often considered a meta-field. The Cybernetics of Cybernetics is, thus, a meta-meta-field. My work is, therefore, a meta-meta-meta-field. (Glanville, 2002, pp. 177-193)

In that spirit, I want to share Ranulph's final lesson to me with you the reader. It was after the conference and after my election to succeed Ranulph as ASC President. We were discussing stridency and the problem of communication in the face of severe disagreement. I forget which of the all too many social or political ills had turned our conversation this way. I showed Ranulph what is known as the Mori Uncanny Valley and related it to the idea of strident disagreement. The context was political and political discourse in America can be viewed as stridency in the making. As Robert Samuelson of the Washington Post put it in 2003 and 2004:

One of today's popular myths is that we've become a more "polarized" society. We're said to be divided increasingly by politics (liberals vs. conservatives), social values (traditionalists vs. modernists), religion (fundamentalists vs. everyone else), race and ethnicity. Today's polarization exists mainly on the public stage among politicians, TV talking heads, columnists and intellectuals. What's actually happened is that our political and media elites have become polarized, and they assume that this is true for everyone else. It isn't. .... For many, stridency is a strategy. The right feeds off the left and the left feeds off the right, ...Polarization serves their interests. Principle and self- promotion blend. (Samuelson, 2003)

Polarization and nastiness are not side effects. They are the game. You feel good about yourself because the other side is so fanatical, misguided, corrupt and dishonest. ... Drab policy debates become sensational showdowns—one side or the other is "destroying" the schools, the environment or the economy. Every investigation aims to expose the other side's depravity ... Politicians, pundits and talking heads all heed the same logic: By appealing to their supporters' strongest passions and prejudices, they elevate their standing. (Samuelson, 2004)

The yelling, the stridency, and the moral clams reach all the way to the heads of our political parties—and this has been the case for generations. Stridency, polarization and labeling seem to create meaning, at least for a moment, as complexity is reduced and decision-making eased. But, at what cost? The reduction of complex problems to polarized labels does not address the complexity of the underlying problem itself. Consensus is not reached because resolution is not the goal; discussants view each other as adversaries rather than seekers of truth or themselves as public servants; compromise is not valued. In today's de-facto context of political debate, objective discourse is marginalized.

> Rejecting an interdependent view of human community invites a deceptive simplification of a conflict by splitting people into separate camps. This "us" versus "them" rhetoric is inherent in any revolutionary viewpoint that seeks to benefit from a class conflict or ideological confrontation. Polarized communication neatly organizes events into contrasting categories, giving the illusion of sharpness of perception, when in reality there is a refusal to gain new insights by listening to the other's viewpoint. (Arnett, 1986, p.34)

Attempts to promote dialogue have traditionally been conducted along one of two paths: the line promoted by the Nobel prize nominee physicist David Bohm—participants in a dialogue must attempt to put aside their partisan differences and enter into a "cooperative space" open to the generation of new ideas; or the line promoted by political realists—where the goal is for each party to compromise and achieve partial victories. Neither of these approaches have been very successful. Despite the asserted good will with which politicians, influencers, and media types supposedly enter a dialogue, all too often the cooperation within a space is limited to the joint agreement to enter it.

The Mori Uncanny Valley is about the rejection of similarity—an expression of cognitive dissonance. It was originally promulgated by a Japanese roboticist Masahiro Mori. Mori's hypothesis was that as the appearance of a robot is made more human, observers' emotional responses to that robot will become increasingly positive, until a point is reached beyond which the response quickly becomes that of strong revulsion. However, as the robot's appearance continues to become less distinguishable from that of a being, the emotional response becomes positive once again.

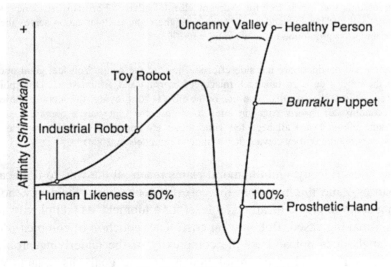

Figure 2: The Original Mori Uncanny Valley

I have extended the Mori Uncanny Valley to the notion of stridency and political disagreement. When we use representations, labels, and category names we have a tendency to demand coherence (a unity or oneness) between the situation, people, process, and so forth to which we are applying the representation and our understanding of the meaning of the representation itself. We do this, consciously or not, in order to ameliorate the risk that our explanations, as well as the actions/ decisions based upon them, are wrong. The Mori hypothesis suggests that once similarity crosses a threshold there can be an emotive reaction which interferes with rational discourse. If there is agreement with the use of representations, then the reaction remains positive. If, however, the observer has either an emotional investment in, or has incorporated into self-identity, a particular representation of the item—a particular representation which is counter to that being expressed, then the articulated label *similar* will produce a negative response.

When one plots emotional response against similarity and claimed identity (see Figure 3), the curve is not a sure, steady upward trend (as would be indicated by a 45 degree sloping line). Instead, there is a peak shortly before one reaches a completely semblant look ... but then a deep chasm plunges below neutrality into a strongly negative response before rebounding to a second peak where the claimed resemblance is complete. When the Mori hypothesis is extended to the realm of scientific perspectives, principles and results, incommensurability is often asserted as the explanation of items found at the bottom of the Uncanny Valley.

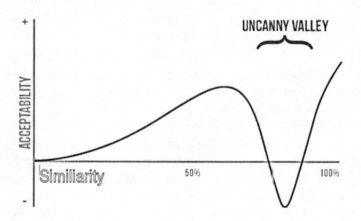

Figure 3 Another View of the Mori Uncanny Valley

The cognitive and communication processes evoked by the Uncanny Valley are shown in Figure 4. Given the existence of two labels (categories) to describe some item/ event/context and the emotional attachment of the observer to the first label, as the perceived characteristics which fall into label #2 increase in similarity the observer shifts his/her perception of that label from 1) surface similarities which only highlight differences to 2) boundary objects, which give rise to explorations of metaphor to 3) emotional opposition, which blocks rational discussion to finally a begrudging acceptance that perhaps label #2 is also a reasonable fit (and thus worthy of being discussed).

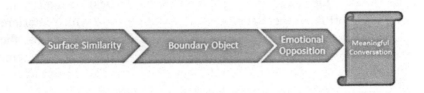

Figure 4: The Progression of Acceptance in the Mori Uncanny Valley

The Uncanny Valley differs from conventional cognition and communication theories which usually assert that the relationship between similarity and acceptance can be plotted at a 45-degree angle. This difference is most obvious when a claim is made that a given representation applies to set of circumstances, situations, individuals, events, and so forth. Once the relationship progresses past surface similarity, it may be claimed that the asserted representation can function as a boundary object (Star & Griesemer, 1989) shared by two competing ideas. Boundary object proponents will claim that the representation allows for the identification of a common area between the two thoughts and that this common are becomes the basis for dialogue. An observer may well react to that claim in a manner approximating Mori's curve rather than the traditional 45-degree line. This often happens when the use of a known label to describe a new set of "facts" is perceived negatively or with regard to the use of a different label for a similar set of circumstances. The Mori hypothesis and the boundary object hypothesis can simultaneously be true where the perceived similarity is at or below the first Mori peak. Thereafter, the hypotheses diverge.

Figure 5 illustrates the three communicative and cognitive regions of our recognition of the Uncanny valley effect. In Area A (where traditional assertions of more information and increased similarity produce higher acceptance) ambiguity reigns—the particularities of meaning are not yet distinct. Only by creating and re-affirming such distinctions do patterns coalesce into defined concepts. The act of drawing these distinctions draws us from Area A to Area B—into the Valley. The definitions and measures we use in Area B (the Valley) are a means of control, and we have both an emotional attachment and reaction to control. By asserting abstract rules, values, and so forth, we can fence off potential conceptual slippages. Mechanically, then, the key to moving from incommensurable stridency (the Valley) to meaningful dialogue is to move the discussion away from the articulated rules and values of Area B—either to the anecdotal evidence of Area A, or the exemplar presentations of Area C.

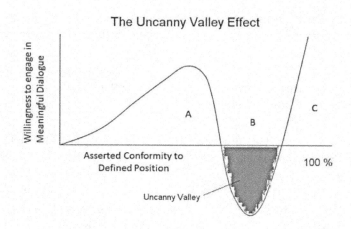

Figure 5: Three Regions of the Uncanny Valley

Figure 5 illustrates the three communicative and cognitive regions of our recognition of the Uncanny valley effect. In Area A (where traditional assertions of more information and increased similarity produce higher acceptance) ambiguity reigns—the particularities of meaning are not yet distinct. Only by creating and re-affirming such distinctions do patterns coalesce into defined concepts. The act of drawing these distinctions draws us from Area A to Area B—into the Valley. The definitions and measures we use in Area B (the Valley) are a means of control, and we have both an emotional attachment and reaction to control. By asserting abstract rules, values, and so forth, we can fence off potential conceptual slippages. Mechanically, then, the key to moving from incommensurable stridency (the Valley) to meaningful dialogue is to move the discussion away from the articulated rules and values of Area B—either to the anecdotal evidence of Area A, or the exemplar presentations of Area C.

By restricting ourselves to a set of representations, we predetermine what might be learned, which will limit the options that appear to be open to us. Stopping at representations is what acts to stifle creativity and ultimately to interfere with effectiveness. What is critical is that the interpretive and retelling efforts NOT stop when the representation gets assigned. To stop at this point is to ignore dialogue and revert to the ascribed coherence and retrospective judgments of identity where the label is the explanation. Such is Area B.

Under the boundary object hypothesis more information and greater similarity will lead to greater acceptance (the 45-degree angle). The boundary object hypothesis and the Uncanny Valley effect will mirror each other in what is shown as Area A. They will diverge greatly in Area B and then re-converge in Area C. If the

Uncanny Valley hypothesis holds for a given representation, then once Area B is reached then three primary possibilities emerge:

1. More information is provided by each of the two competing ideas in support of their own respective positions—incommensurability will ultimately be asserted due to the lack of common dialogue.
2. The two positions are abstracted back to a level of greater generality (i.e., less information is asserted and greater abstractions replace the missing information) —this has the opportunity of moving the dialogue to the left—and back into Area A where boundary objects can be found.
3. Enabling constraints are asserted which restrict the context and applicability of the representation ceteris paribas. By restricting the context, the dialogue can be moved to Area C.

It was Ranulph's inspiration that allowed me to see the possibility of movements described by numbers 2 and 3 above. I regard these as the cybernetic approach to escaping stridency.

I close with a quote from Ernst von Glasersfeld:

> Cybernetics is the art of creating equilibrium in a world of possibilities and constraints. This is not just a romantic description; it portrays the new way of thinking quite accurately. Cybernetics differs from the traditional scientific procedure, because it does not try to explain phenomena by searching for their causes, but rather by specifying the constraints that determine the direction of their development. (Glasersfeld, 2010, p. 136)

Ranulph helped me to better see constraints and in doing so helped open my world to many new possibilities. I shall be forever grateful.

References

Arnett, R. (1986). *Communication and Community: Implications of Martin Buber's Dialogue*, Carbondale, IL: SIU Press.

Churchill, W., & Churchill, W. S. (2003). *Never given in!: The best of Winston Churchill's speeches*. New York: Hyperion.

Glanville, R. (2002). Second-order cybernetics. Invited chapter for *Encyclopedia of Life Support Systems*, UNESCO. Retrieved March 3, 2014 from www.eolss.net/ . (Reprinted as chapter 1.11 of Glanville, R., 2012, *Cybernetic Circles, The black box, Vol. 1*, echoraum-WISDOM, Vienna.)

Lissack, M. (in press). "What Second-order science reveals about scientific claims: Incommensurability, doubt, and a lack of explication. *Foundations of Science*.

Lissack, M. (2015). Restoring dialogue to political debate: A Buberian approach. Unpublished paper.

Mori, M. (1970). The uncanny valley. *Energy, 7*(4), 33-35

NBC News "Meet the Press." (2005, May 22). Howard Dean interview. Retrieved November 6, 2015 from http://www.nbcnews.com/id/7924139/ns/meet_the_press/t/transcript-may/#.Vj9Ensqr0Q0

Samuelson, R. (2003, December 4). The myth of a polarized America. *Newsweek*. Retrieved November 15, 2015 from http://www.realdemocracy.com/polrmyth.htm

Samuelson, R. (2004, June 30). How polarization sells. *The Washington Post*. Retrieved November 15, 2015 from https://www.washingtonpost.com/archive/opinions/2004/06/30/how-polarization-sells/56b6e027-d1fe-4707-96a0-c442255a6ed1/

Star, S., & Griesemer, J. (1989). Institutional ecology, "translations" and boundary objects: Amateurs and professionals in Berkeley's Museum of Vertebrate Zoology, 1907-39. *Social Studies of Science, 19*(3), 387-420 von Glasserfeld, E. (2010). *Partial memories: Sketches from an improbable life* (p. 136). Exeter, UK: Imprint Academic.

Bunnell, P. (2010). *Reflection Refraction.* Un-retouched photograph.

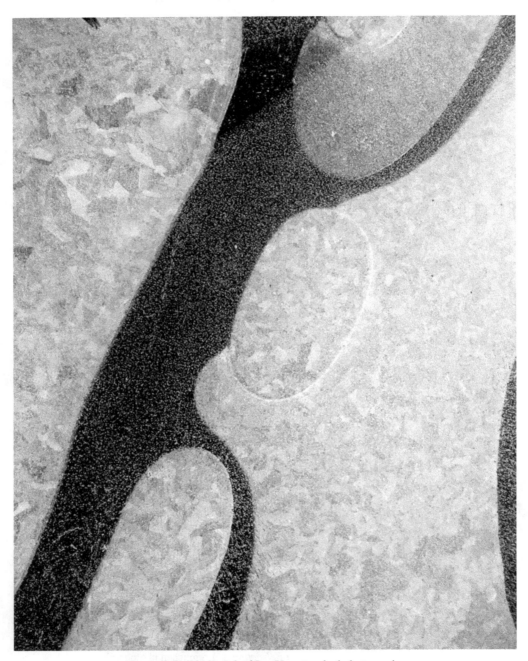

Bunnell, P. (2010). *Inlaid Ice*. Un-retouched photograph.

Cybernetics and Human Knowing. Vol. 22 (2015), nos. 2-3, pp. 131-144

Designing Together

Thomas Fischer[1]

During our joint years on the executive board of the American Society for Cybernetics, I had the opportunity to design together with Ranulph Glanville, the cybernetician and design researcher who established and drew attention to the value and structure of designing together. In this paper I outline the context of our collaboration as well as aspects of Ranulph's design theory, followed by three short email exchanges that indicate the nature of our online design conversations. The purpose is to offer a glimpse of data recorded in Ranulph's design practice, complementing the theoretical work Ranulph has published on the subject of designing, and to offer a verbatim insight into Ranulph's working style. I point out that on a day-to-day basis our design conversations were subject to various challenges, and that they were bleak at times. I point out that, nonetheless, the thinking and ethos behind our work, and the match of form and content made our work more than worthwhile.

Introduction

Today's leading edge of design theory being cybernetic, and today's leading edge of cybernetics being design cybernetics (Fischer, 2014) is due, first and foremost, to the work of Ranulph Glanville (1946–2014), and to his self-immersion in his designerly cybernetics, or cybernetic designing. Ranulph Glanville's work was, both in terms of his acting and in terms of the theories he produced, centered on togetherness. When I first met Ranulph in 1999 at a colleague's family dinner party in Geelong, Australia, he did not strike me as someone with much interest in togetherness. I was introduced to Ranulph as having done a PhD with Gordon Pask, as a leading mind in digital design theory, and as published extensively. I asked Ranulph whether he had an *Erdös number*—the measure of collaborative distance between academic co-authors and the Hungarian mathematician Paul Erdös (Goffman, 1969). Ranulph responded in a somewhat off-putting way to the effect that he didn't co-author as a matter of principle, because thinking is personal. I met him again a few years later when he was a panelist at my semi-annual postgraduate research reviews at RMIT in Melbourne. Repeatedly, his responses to my work were brilliant, unforgiving and, at times, again off-putting. A few years later I found out that Ranulph, on occasion, did produce co-authored academic papers. There was one (Glanville & Varela, 1981), which he was particularly fond of (Glanville, 2002). And eventually, there was one that I am fond of (Fischer & Glanville, 2011).

Those prepared to tolerate Ranulph's characteristic bluntness and open to what he had to offer, could benefit greatly from his insights and generosity. As I learned to listen to him, he became an invaluable, unofficial, although de-facto advisor to me. In 2008 I followed Ranulph's invitation to join the American Society for Cybernetics (ASC), with which he had been associated for several decades. The ASC has been the

---

1. Department of Architecture, Xi'an Jiaotong–Liverpool University, China. Email: Thomas.Fischer@xjtlu.edu.cn

spiritual home in particular of second-order cybernetics since 1964, organizing regular conferences on cybernetics and maintaining close ties to key periodicals in the field. Soon after I joined—in 2009—Ranulph was elected as the ASC's president, and I was elected as the secretary of the executive board. We served in these roles for two terms until Ranulph passed away in late 2014.

After I obtained my PhD, besides our joint paper mentioned above, besides occasional comments I wrenched from him on my writings, besides a joint visit to the National Museum of the US Air Force and Wright-Patterson Air Force Base in Dayton, Ohio together with Ranulph's wife Aartje en route to the ASC's 2011 conference, and besides a visit he paid to my university in China in early 2012, virtually all of our many joint efforts were ASC related. This provided me with the unique opportunity to design together with the cybernetician and design researcher who like no other established and drew attention to the value and structure of designing together.

Those who met Ranulph, who attended his talks or read his writings, and especially those whom he offered academic feedback, are aware of his keen insights into a wide range of human forms of expression. These included, amongst others, music, theatre, architecture, technology, science, philosophy and education. He often mentioned works he was passionate about such as La Monte Young's and Marian Zazeela's *Dream House*, John Cage's *4'33*, and the SR71 Black Bird reconnaissance aircraft designed by Kelly Johnson; and he often referred to artists he admired including Laurie Anderson, Samuel Beckett, Captain Beefheart, and David Bowie. There were many more. Ranulph's pet references have in common that they are subtle and sophisticated in raw, artistic ways, on the forefront of their respective fields, and that in one way or another they all relate to (his) design cybernetics.

Designing was a central theme in Ranulph's thought, in his writings and in his everyday life. Ranulph described designing as a subjective conversational practice of a "self" with an (imagined or actual, human or not) "other." Taking the form of a feedback loop of acting and understanding, one purpose of this "conversational cycle" (Glanville, 1999, p. 88) is to get out of control so as to allow this conversational process to arrive at something that is new, at least subjectively. This contrasts starkly with our prevailing cultures' fixation on systems that are to be in control and directed towards known, specified goals. This perspective of Ranulph's, with its counter-intuitive valuing of ignorance (Glanville, 2009a, p. 160ff.), of misunderstanding and of error (Glanville, 1997a, 2007a), will probably require decades to be grasped and appreciated more broadly. To Ranulph, designing was not a temporary means towards practical ends. Instead, he was rather suspicious of utilitarian in-order-to-type modes of operation. To him, design was a verb, a continuous way to practice and to celebrate his ethical position, and integral to his identity. Designing was the form and its ultimate criterion was delight. These were primary. The content and its goals could be just about anything, and, to Ranulph, were often secondary. This is not to say that in Ranulph's relationships with others, content and goals of designing did not matter, since these were often what drew others into illuminating and productive conversations with him. But it was the process of acting together through which Ranulph defined himself and his relationships with others.

Ranulph clarified, and significantly expanded upon conversation theory (Pask, 1976), developed by his mentor and PhD advisor Gordon Pask. Accordingly, he saw novelty as arising from interaction in conversation, and "betweenness [as] the source of interaction and is also its mode and its site" (Glanville, 2000, p. 4). His favorite metaphor to relate the circular process of designing to the outcomes of design was that of a wheel leaving tracks in the sand (Glanville, 1997b, p. 4). Prioritizing interaction over the attainment of pre-conceived goals, Ranulph was more interested in the continued turning of epistemological wheels than in reaching known destinations, and, therefore, saw the *good enough* as better than the *best*.

Ranulph's attitude towards art, as he saw it, was yet more fundamental than his absorption in designing. In his notion of art, neither is form subordinate to content, nor is content subordinate to form. Ranulph saw art as the match of content and form: A figure that does what it is. This was his essential benchmark criterion in any form of expression, and, I believe, his perpetual aspiration. He did not see this as a necessary goal to attain through intentional control. When a project failed to arrive there, he saw it as a practice run necessary to get better next time (Glanville, 2007b, p. 1191). Often, his projects did arrive at the match of content and form; and often, he helped others see their projects arrive there. When they did, it filled Ranulph with delight.

Ranulph described how he first came to appreciate the match between form and content during a talk Heinz von Foerster gave when visiting Brunel University in 1972. In this talk Ranulph recognized

> the art that, at many levels, matches the form and the content to make something very special, beyond the bounds required of science, able to (be taken to) stand on its own two feet and to provide us with endless opportunity to wonder and delight, knowing we have no alternative but to make our own meanings. (Glanville, 1996, p. 277)

In another paper he notes:

> I believed (and still do) the form and the content of works of art [map onto each other]: the form indicated the content, the content required the form ... Coming from a background in architecture and music, this meant a lot to me (and still does: cybernetics is my art). (Glanville, 2009b, pp. 130-131)

During our joint years in the ASC, we executive board members produced communications, images, logos, posters, flyers, events, procedures, services, publications, and so forth. This was a group effort, with a sometimes-changing cast of five or six officers, plus trustees, conference hosts, past officers, auxiliary working groups, and other volunteers. Ranulph was a hub between many diligent, oftentimes collaborating spokes. Together, we were co-editors, co-organizers, moderators and collaborators. But to Ranulph and me in particular, the basic paradigm of our collaboration was design. Design was our shared background, it was what brought us together, and at the same time it was a key concern of the society we were now working for.

Not surprisingly from a design perspective, our administrative work often seemed lossy and inefficient from within. Visible achievements we made were mere tips of icebergs, with significant efforts going mostly unnoticed beneath the surface. This included unsuccessful efforts to establish non-U.S.-based, non-U.S. citizens as ASC treasurers with control over U.S.-based bank accounts in the post-9-11 climate; the restoration of the ASC's legal standing with the establishment of a required board of directors and a complete overhaul of the Society's by-laws with considerable help and advice from Randall Whitaker and Klaus Krippendorff, and elaborate online consultation of the entire membership; and the successful collection of nearly USD10,000 which were fraudulently withheld by an online payment service we used, with the generous and competent support of the ASC's legal advisor Ern Reynolds.

While the discrepancies between our efforts and effective forward-movement may not have been out of the ordinary for volunteer-based administrative work, they surely must be experienced to be fully appreciated. The bottom line accomplishments of Ranulph's two administrative terms for the ASC have been summarized, with a broad brush, in annual (vice) president's reports and in the citation of the special awards Ranulph and his wife Aartje Hulstein received from the ASC during the last year of Ranulph's presidency (ASC, 2014). The work and generosity that went into these accomplishments were much more significant than what became apparent.

More often than not, our design conversations had Ranulph, having been involved in cybernetics and the ASC for a much longer time, on the understanding arc, and me, being more fluent in parts of the digital toolbox, on the acting arc of the design-conversational cycle. I never stopped seeing Ranulph, over 26 years older than me, as my mentor. Nonetheless, this role allocation soon softened and gave way for a design relationship without hierarchy. Ranulph never gave orders or made definitive requests. Similar to the way Alcoholics Anonymous operate, Ranulph collaborated "based on attraction rather than promotion" (Glanville, 2013, p. 193), and in accord with von Foerster's preference of internal ethics over external morals (von Foerster, 2003). Accordingly, he could rely on us, inspired by his example, to push ourselves. At the beginning of his first term, Ranulph asked us other officers to revise their election statements into descriptions of our roles and ambitions. In this way, we formulated our own orders and Ranulph would have had something to refer to in case we would not deliver. As far as I know he never had to; and he dropped this practice in his second term.

Despite limited resources and despite other commitments, despite conflicting expectations of others, and despite our differences in working styles and levels of experience, we collaborated with remarkable discipline and patience. Without ever making this explicit, Ranulph and I neither spoke nor acted in ways that would add to any pressures or frustrations we might have been going through inside or outside of the ASC. In this way, we protected our limited resources and kept ourselves going despite sometimes-low energy levels. Instead, we challenged each other gently by asking each other questions, by making careful suggestions, and sometimes by making alternative proposals. This was a departure from our earlier interactions when we were playing different roles. When I was a student and he was my reviewer, the work was my responsibility, and he was an outsider who criticized it when his standards were not met; and he sent me back to the drawing board to start over, regardless of how frustrating this was. Now, we were both insiders in joint endeavors, with shared goals, and shared resources. We did not hide when we thought something could be improved, but this never involved negativity, protest or vetoes. We both understood well that designing goes on until a critically important resource runs out, and we made sure that it would always be time that ran out, and never our care for our projects or our patience for each other. In this way, we almost always satisfied the tasks at hand, while maintaining the working chemistry between us. This was important because the next handful of tasks was usually already underway or just around the corner, and sustained intense volunteer work depends on positive working chemistry. We managed, as Ranulph might have put it, without upsetting the apple cart.

While we designed together as equals, we were, in other ways, clearly not equal. It was Ranulph who kept numerous internal and external projects on track, who remembered all the important birthdays, who also reached out to other groups, and who wrote hundreds, or possibly thousands of thank-you emails for smaller and larger gestures of support we received. This process was subject, amongst other challenges, to time pressures, other work and family commitments, poor health, limited resources, Internet blockage in China, technical difficulties and that they were dreary at times.

Our Design Conversations

In this section, I offer a glimpse into a few of the design interactions Ranulph and I had while working for the ASC. These are chosen somewhat randomly out of a vast archive of collaborative data and working communications, and cannot capture in full the impact of the challenges listed above. Dates are shown as recoded in the email exchanges. Mostly due to time zone differences between our changing locations they may not reflect the timing of communications accurately. Multiple other exchanges between Ranulph and myself have been going on simultaneously, through various channels. Some typing mistakes have also been fixed in the messages shown below.

Logo and Diagram for the 2013 ASC Conference
Ranulph (25-02-2013): "I believe you have everything else you need, except perhaps a logo. How about a Möbius in a sea of learning, with understanding on one 'side' and action on the other."
Tom (25-02-2013): "Here is a first draft of that mobius strip [second image in the top row of figure 1 below]. We're not sure about the sea of learning."

Ranulph (26-02-2013): "You are quick! Let me suggest that action is written upside down and from right to left following on from 'ing'. I'd like to see 'tanding' all written slightly bigger. I hope this makes sense. But as an idea, it seems to me this works rather well. As for the 'learning' element, here's an idea. write it along the edge all the way round, like a small trimming."

Figure 1: Successive stages in the 2013 conference logo design development.

On 03-03-2013 an executive board meeting was held via Skype. It included a short President's report on the development of the 2013 conference website, promotion activities and fee structure, and a longer discussion on ideas for moving the ASC and cybernetics forward.

Ranulph (04-03-2013): "I am being sent to hospital. I probably won't be in long. I've had a terrible cough, cold etc., and they think it may be to do with my heart. So I won't be fielding things. Please communicate amongst yourselves and not through me."

Ranulph (05-03-2013): "I was allowed home from hospital under very strict instructions that I should return instantly if there was any deterioration. The diagnosis was that I don't have a heart condition, which is a blessing. I do however have lung damage from a very serious childhood disease and that renders me susceptible to bronchitis. I have that, a heavy and particularly nasty cold, and on top of my rundownness, this means I have been instructed not to work or travel by the hospital consultants as well as my GP."

I went on completing the logo without Ranulph's feedback, relying however on some verbal comments Candy made. For scale consistency, "Learning" moved from its subordinate position in the trimming onto the Möbius strip, alongside "Understanding" and "Action", the latter having changed to its present continuous form "Acting" for consistency and to stress its being a verb. The word "Understanding" turned upside-down as per Ranulph's suggestion.

At some point Ranulph circulated the following diagram, which he has designed, and which he explained thus: "Acting (A) leading to Understanding (U), and vice versa, through Learning (L), presented first through two linear processes and then as a circular process as an integrated whole."

Figure 2: Diagram describing the relationship between Acting, Learning and Understanding, designed by Ranulph.

Ranulph (30-04-2013): "Are you ok with the little diagram I put together?"

Tom (30-04-2013): "Of course I am OK with the diagram and we should use it as you see fits."

Ranulph (30-04-2013): "Should we put that diagram at various places on the web site? Or bits of it, for instance in the description of the process?"

At that time, I saw this new diagram as potentially redundant and interfering with the conference logo, while Ranulph saw it as offering a helpful framing. I argued that the recognition of this circular relationship could be expected to emerge from conversations at the conference while Ranulph saw it as a starting point from which conversations should depart. Eventually, we used his diagram to illustrate the descriptions of conference's premise and main conference activities on the conference website (ASC, 2013). As Ranulph had anticipated, it worked as a useful point of departure at the conference, and nobody seemed to have been confused by its existence next o the conference logo.

The ASC's i2s Conference Poster

In early August of 2013 an invitation reached the ASC from the First Global Conference on Research Integration and Implementation to submit a digital poster for display at the event in Canberra between the 8th and 11th of September. Ranulph emailed to the Executive Board and some other ASC members:

Ranulph (05 Aug 2013): "Does anyone feel they can do anything about this? Obviously, it would at worst do no harm and at best might help. But I know we are all busy and most of us are very tired. If you have an idea or a suggestion for someone who might have, please send it round to us all! This needs quick action."

Nothing happened and Ranulph raised this matter again during an Executive Board Skype meeting on August 31st. He suggested putting together a poster based on the diagram he developed for the 2013 conference (shown in figure 2 above). I agreed to collaborate on this with him.

Ranulph (02 Sept 2013): "How's the poster. If you've no idea, I thought that simply placing the 3 different visual components above each other (with the maths signs + and = between). The short first para of text goes on top (bigger letters), the other paras go between the diagram visuals, and the ASC info at the bottom. I was thinking of greying out the text a bit. But I'm sure you'll have thought of something good."

Thomas (02 Sept 2013): "I made the attached before I read your email. It is just a draft and the P(ractice) and T(heory) in the formula are not yet explained or contextualised properly. I changed your spelling to American. Please let me know what you suggest and I will be happy to execute as far as possible."

Ranulph (02 Sept 2013): "It's not that I have a fixed idea. I had one, in case there were no others. And one you might consider. That's all. I added an extra paragraph (short) emphasising circular causality, to be an attention grabber. I am a bit concerned about your choice of image. I know you like it. For me, however, it can easily be understood as trivial and silly by those who were not there and have no idea what we're doing. It looks like some sort of encounter group! So I'd not use it. I'll try sometime during the day to sketch what I had in mind, but I thought I'd managed fairly well in words. I just don't have the right stuff for this on this computer and I've a mountain of dissertations to read in Glasgow, for exams tomorrow. You can choose to use what I send, or not!"

Ranulph then asked his wife Aartje for comments and forwarded her response to me.

Aartje (02 Sept 2013): "I think the poster would be more powerful if you did not have text through the diagram and used words rather than the letters, or at least gave explanations to the letters in the diagram. You also said at some point you wanted to use theory and practice in there.

The logo would be better at the bottom of the page with the details of the ASC next to it. Then the first sentence is more of an eye catcher. Do you use continue to exploring as proper English, to me it sounds odd."

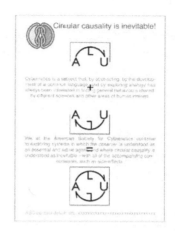

Circular causality is inevitable!

+

=

Cybernetics is a subject that, by abstracting, by the development of a common language, and by exploring analogy has always been interested in finding general behaviors shared by different sciences and other areas of human interest.

We at the American Society for Cybernetics continue to explore systems in which the observer is understood as an essential and active agent, and where circular causality is understood as inevitable—with all of the accompanying concomitants, such as side-effects.

American Society for Cybernetics
secretary@asc-cybernetics.org
www.asc-cybernetics.org
http://www.linkedin.com/groups?gid=2206885

Circular causality is unavoidable!

+

=

Cybernetics is a subject that, by abstracting, by the development of a common language, and by exploring analogy, has always been interested in finding general behaviors shared by different sciences and other areas of human interest.

We at the American Society for Cybernetics continue to explore systems in which the observer is understood as an essential and active agent, and where circular causality is understood as inevitable—with all of the accompanying concomitants, such as side-effects.

Circular causality is unavoidable!

+

=

Cybernetics is a subject that, by abstracting, by the development of a common language, and by exploring analogy, has always been interested in finding general behaviors shared by different sciences and other areas of human interest.

We at the American Society for Cybernetics continue to explore systems in which the observer is understood as an essential and active agent, and where circular causality is understood as inevitable—with all of the accompanying concomitants, such as side-effects.

American Society for Cybernetics
asc-president@asc-cybernetics.org
http://www.asc-cybernetics.org
http://www.linkedin.com/groups?gid=2206885

Figure 3: Successive stages in the ASC's i2s conference poster design development.

Thomas (03 Sept 2013): "I didn't find that added paragraph. Where? Attached is a revised version, with what I believe is the format and aspect ration they expect (they are quite blurry about this, unless I am missing something). I am still not sure how to explain the letters T and P in the formula/diagram. I am slow now, overloaded (semester started yesterday)."

Ranulph (03 Sept 2013): "It's the first big paragraph I already sent you both as revised text, and in my sketch. The image I'm looking at is the one that has appeared twice in Dropbox. It has the photo. I'll look again later. Sorry you are so loaded. I sort of know how it feels! Good luck and thanks! I have to go to examine."

Ranulph (04 Sept 2013): "I attach the draft I received most recently. It's fine by me, if you're happy with it. One with a photo would be good, too—I'm not just sure of the photo you chose, for the reasons I gave. Looking at the text, I think unavoidable would be better than inevitable. Could you change that?"

Thomas (04 Sept 2013): "Please find a revised version attached. T and P are now THEORY and PRACTICE. Inevitable is unavoidable, and the links are hot. I also added a comma between 'analogy' and 'has' in the first paragraph - please check if that's OK. We have less silly ASC photos, but this super simple layout makes it difficult to integrate a photo nicely. I wonder if no photo is better, but will be happy to try out suggestions if you want me to."

Ranulph (04 Sept 2013): "What I sent was a sketch that I roughed out in case it was needed. I didn't mean to impose it. I think it has an advantage: it is very clear and strong, in its starkness. But I think what you did was also interesting and eye-catching: my problem was only the choice of photo, for the reasons I explained, and a couple of layout matters and text change matters, which are easily corrected. So I'm happy with either basic poster, but I don't want to add any more to your load. The draft you sent me is just fine. All I would still consider is the alignment of the text and the diagrams. I think that the top of the first paragraph and the bottom of the second should align with the diagrams, rather than appearing to be slightly out of alignment. However, this may just be my screen. If you're happy with this poster, please just submit it. If you rather use your photo one, please submit it. I am happy to leave it as your choice, unless you want me to make that decision."

The 2014 Conference Logo

In early December of 2013 the theme for the 2014 conference was finalized with a tweak from "Living Change" to "Living in Cybernetics". With this decision, we were ready to start working on promotional material and the website, which again started with the design of artwork and visual identity. Key stages of the design development for the conference logo are shown in figure 4 below.

Thomas (08 Dec 2013): "Attached are some sketches for conference graphics [left of top row in figure 4] for your comments."

Ranulph (08 Dec 2013): "What your images bring to mind is knitting, crocheting or lace making. I think the problem for me with your suggestions is that they suggest a mess, whereas we could have a bringing together producing something of beauty, and a pattern. What do you think?"

Thomas (09 Dec 2013): "How about these [centre and right of top row in figure 4]?"

Ranulph (09 Dec 2013): "Thank you for the effort you always put into the ASC. Your contribution is enormous. So thank you for the graphics you are developing for the next conference. I think this latest model is a bit 50's, and I'm not sure I see the graphic story. The earlier one suggested local chaos within the general frame of order. This is rather tidy, and suggests a seashell. What I was hoping was for something that would show the doing together of different strands that, through their interaction, produce something with pattern and structure. This is exactly what knitting does, which is why that come to my mind as an appropriate graphic symbol. It's also closely related to your ASC logo. Would you mind giving this a bit of a rethink?"

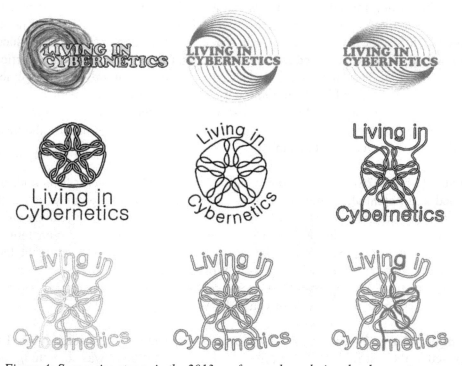

Figure 4: Successive stages in the 2013 conference logo design development.

Thomas (13 Dec 2013): "I find the images of knitting and crochet challenging with regards to creating a coherent, self-contained logo without too much detail. Here is a new idea based on your and Candy's input and suggestions. The 5-fold symmetry is a reference to the ASC's 5 decades. What are your thoughts?"

Ranulph (13 Dec 2013): "This must have been difficult to produce, though I see it as a sort of variant of the ASC logo. What worries me is that it's closed. It suggests a complete and perfect system. I had in mind that what happens in knitting is that several strands come together and form a stable whole through their interaction, even though there is no knot at all. I'd imagined this as happening between strands, which are left open each end, but knit together in the middle. I don't have the software or the skill to try to show you what I had in mind, as a possible inspiration, but I'll see if I can draw something and send a photo. Thank you for your efforts."

Thomas (13 Dec 2013): "Attached is a concept for the circular knot as an open system, integrated with the typography. This is incomplete, showing the principle only, because it takes a bit of work to do well and completely."

Ranulph (14 Dec 2013): "Sorry I didn't reply earlier. I am both busy and allowing myself to be a bit lazy here in Tallinn. I begin to like where this is going. I think the idea of the writing forming two threads that are knitted together to give the pattern is excellent. I think, therefore, cybernetics will have to be split into cyber and netics. I would like to see the text a bit removed in space from the "flower" (you've produced a rose), with the letters connected as if on a thread, and also placed essentially vertically above and below. Then there are 2 threads (living and in) that are knitted together to make the flower, and which then come out as 2 threads cyber and netics. There are now several metaphors, of which the most powerful is the beautiful emergent whole (the rose), which occurs when we live in cybernetics (when we compose the threads together). A very nice visual metaphor."

Thomas (14 Dec 2013): "Yes, your email is very helpful. However, it is both quite detailed and potentially ambiguous. I am sure I can only disappoint your concrete expectations - unless you are a little bit flexible. Putting successive letters onto one single piece of string will be clumsy. A primary-school type handwriting font may allow this, but will likely be too messy for a logo. I was thinking of keeping the letters separate, but having multiple strings (say 8 or so) run into the rose from the letters from all directions. In this way the system would be 'open'. Words on a string would close the system again. I hope I got you right. What do you think?"

Ranulph (15 Dec 2013): "I don't have contrite expectations, just some ideas. I was surprised at the last figure you sent me, and then thought there was something rather good in it, which I'd certainly not thought of. So I rethought everything, as designers do, and tried to rethink the idea I'd half had. So your suggestion seems to me worth having a go at, if you're ok with it. But I'd try not to be too determined about keeping the 5fold symmetry."

Thomas (15 Dec 2013): "Here is a new draft. I forgot to increase the spacing between the rose and the letters. What do you think?"

Ranulph (15 Dec 2013): "I think this is almost perfect: it's better than what I had in mind, a good mix of what you thought and what I did. What I would do is remove the lettering from the rose. Make the space between them much bigger, so the threads going into the flower can be seen. I might also experiment with having the lettering in horizontal straight lines, to help make the thread going in to knit the flower more immediately visible. It also relates nicely to the ASC logo, of course. I'm not sure about the colour."

Over the following three days we fine-tuned the design, making small choices regarding which letters should be connected, and which thread should be the one highlighted in red colour. This was settled on the 18th with a reference to Ranulph's initials, which I suggested as a joke, but he liked it and went along with it:

Thomas (18 Dec 2013): "We should have connected the r and the g!"

Ranulph (18 Dec 2013): "It's not too late, and it's a nice joke!"

   Probably still thinking about Ranulph's 30-04-2013 suggestion to use the elements of his diagram shown in figure 2 above in various places across the conference 2013 website, we went on to use the four threads that run through the resulting logo to illustrate the four main elements of the conference on the 2014 conference website as well as in the conference booklet (edited by Christiane Herr): One n-t thread became the symbol for "Cybernetics in the Past," the i-b thread became the symbol for "Cybernetics in the Present," the other n-t thread became the symbol for "Cybernetics in the Future," and the r-g thread became the symbol for the main theme "Living in Cybernetics,"

Observations

I had the valuable opportunity to learn about Ranulph's design thinking not only by reading his writing, but also by being a collaborator in his design practice. I experienced our joint designing as a continuous attempt to produce straightforward and appealing circumstances and resources out of circumstances and resources that were not always straightforward and appealing. Welcoming me into his cybernetic circles, Ranulph showed me how in a world of obstacles and challenges, he made theory and models about the beauty and ethics of engaging with this world, and thereby made it beautiful and worthwhile for himself (and for me). I experienced and remember this collaborative process as a great privilege, designing together with the cybernetician and design researcher who established and drew attention to the value and structure of designing together. We found shared understanding between us, accepted that the perfect is the enemy of the good-enough, interacting as a generous, circularly-causal wheel, leaving various tracks in the sand. Ranulph who had initially introduced himself to me suggesting he would not collaborate eventually turned out to be the best collaborator imaginable. A figure who enacted what he thought, and who masterfully explained his acting.

Acknowledgments

I gratefully acknowledge the valuable comments I received on drafts of this paper from Aartje Hulstein and Albert Müller, and their kind permission to publish the above email exchanges.

References

ASC. (2013). *Description of the Main Part of the 2013 ASC Conference.* URL: http://asc-cybernetics.org/2013/?page_id=403 [accessed 25-08-2015].

Fischer, T., & Glanville, R. (2011). Besides designing to interact: Interacting to design. In *Delight and responsibility: Proceedings of the 2011 International Conference on Interaction Design, School of Design*, Hong Kong: The Hong Kong Polytechnic University.

Fischer, T. (2015), Wiener's prefiguring of a cybernetic design theory. *IEEE Technology and Society Magazine, 34*(3), 52–59.

Glanville, R., & Varela, F. (1981), Your inside is out and your outside is in. In G. Lasker (Ed.), *Applied systems & cybernetics, Vol II* (pp. 638–641). Oxford: Pergamon.

Glanville, R. (1996), Heinz von Foerster: The form and the content. *Systems Research, 13*(3), 271–278.

Glanville, R. (1997a). The value when cybernetics is added to CAAD. In K. Nys, T. Provoost, J. Verbeke, & J. Verleye (Eds.), *The added value of computer aided design. AVOCAAD First International Conference* (pp. 39-56). Brussels: HWK Sint-Lucas.

Glanville, R. (1997b). *A ship without a rudder.* In *The black box: Vol. I. Cybernetic circles* (pp. 125-135). Vienna: edition echoraum.

Glanville, R. (1998). A (Cybernetic) Musing: The gestation of second order cybernetics, 1968–1975—A personal account, *Cybernetics and Human Knowing, 5*(2), 85–95.

Glanville, R. (1999). Researching design and designing research. *Design Issues, 15*(2), 80–91.

Glanville, R. (2000). *The value of being unmanageable: Variety and creativity in cyberspace.* In *The black box: Vol. I. Cybernetic circles* (pp. 521-531). Vienna: edition echoraum.

Glanville, R. (2002). Francisco Varela (1946–2001): A working memory. *Cybernetics and Human Knowing, 9*(2), 67–76.

Glanville, R; (2007a). Grounding difference. In A. Mueller, & K. H. Mueller (Eds.), *An unfinished revolution? Heinz von Foerster and the Biological Computer Laboratory 1958–1976* (pp. 361–406). Vienna: edition echoraum.

Glanville, R. (2007b). Try again. Fail again. Fail better: The cybernetics in design and the design in cybernetics, *Kybernetes, 36*(9/10), 1173–1206.

Glanville, R. (2009a). Black boxes. *Cybernetics and Human Knowing, 16*(1, 2), 153–167.

Glanville, R. (2009b). A (cybernetic) musing: The gestation of second-order cybernetics. In *The black box :Vol. III. 39 Steps* (pp. 125-137). Vienna: edition echoraum.

Glanville, R. (2013). A (Cybernetic) Musing: Anarchy, alcoholics anonymous and cybernetics, Chapter 1. *Cybernetics and Human Knowing, 20*(3/4), 191–200.

Goffman, C. (1969). And what is your Erdös number? *The American Mathematical Monthly, 76*(7), 791.

Pask, G. (1976). *Conversation theory: Applications in education and epistemology.* Amsterdam: Elsevier.

Bonding is a heart conversation . . .

I am a wave that flies
I am a wing that flows

I breathe wave wings
through the centre of myself

Between the heart beat
and the breath

Deep in the time of distinctions. . .
The space
among the cells

Between the lace webs
of light that bond
cell to cell
in relations of harmony

In the moment of conception

We perceive
the conversation...

Bunnell, P. (2010). *Red Bridge Reflection Reflexion*. Un-retouched photograph.

Cybernetics and Human Knowing. Vol. 22 (2015), nos. 2-3, pp. 147-154

Glanville's Consistency

Philip Baron[1]

The late Gary Boyd provided a critique of Glanville's work saying it lacked coherence and a unified position. Although Glanville did provide his own argument to Boyd's critique, there is an additional aspect that is overlooked, yet it is so obvious. Apart from Glanville's mammoth contribution to cybernetics both in his writings and his academic activities, there exists a theme that I believe should be celebrated. In this paper, an argument has been put forward for what may be at least one of the unified themes of Glanville's contribution to second-order cybernetics. I dedicate this paper to a lost teacher and mentor. This paper departs from a traditional academic format.

Introduction

Glanville recounts a conversation he had with the late Gary Boyd: "The late Gary Boyd once said to me that he could see no coherence in my work. That there was no theme and there was no position: my worked lacked a theory" (Glanville, 2012, p. 19). Glanville (2012) responded with an account on how cybernetics tends to be multidisciplinary, weaving across disciplines, while also being a meta discipline. He describes how many cogs get set in motion with each cog important in its own right. His full retort was an interesting philosophical answer, but it seemed a justification rather than an adequate response. Having worked with Glanville and read much of his works, I did not agree with Boyd's critique, but neither with Glanville's answer. It is true that Glanville has written on various topics across several domains, but there is a straight forward central aspect tying all of his seemingly divergent works together. I propose what I observe and experience as Glanville's legacy in cybernetics of cybernetics: consistency.

I met Ranulph at an American Society for Cybernetics (ASC) conference in July 2011. He came to speak to me at the conference dinner and cajoled me into getting more involved with the ASC. He took a persuasive stance as he saw that I did not enjoy the conference. Having shared some correspondence following the conference, I decided to attend the 2012 conference where he asked me to be part of the organising committee. Following this he invited me to be part of the editing team for the ASC's proceedings for 2012, 2013 and 2014. From the end of 2012, I was communicating with Ranulph extensively. I counted 1319 emails sent between Ranulph and myself between 2012 and December 2014. Apart from the editing work, there were many conversations about other topics. At the time of writing, myself and a colleague have been organising Ranulph's publication repertoire, which exceeds 400 units (all forms—print and video—to be linked to his new website). Sadly this mammoth work contribution tells only part of his story.

---

1. Email: pbaron@uj.ac.za

While working on the ASC conference proceedings for the 2014 conference, Ranulph's cancer started to get worse. He indicated this to me in emails with his last prognosis saying that he may be out of sorts and probably having to reduce his efforts even in the editing tasks. Being the administrator of the conference email account, I decided to exclude him from the group emails and discussions that we were sending each other as the guest editors for this journal (4 members). After about 10 days I received an email from Ranulph asking what is happening with the proceedings and reminding me of things that still needed to be attended to. I told Ranulph that we were attending to those items and that I had just shielded him from the many emails to give him space, and to not bother him with these trivial editing issues. He was almost offended, so I immediately put things back to the way they were and Ranulph was back in the loop. He said that he did not need to be left out owing to his illness, but just wouldn't be reliable in performing tasks. Ten days later he passed away. I was notified by the people close to him, but it was not a surprise as I had not received an email from Ranulph in the week. A lapse in Ranulph's consistency meant something went wrong. Ranulph had an exceptional work ethic. His reliable emailing will be sorely missed by those he worked with.

Conversation and the Correct Use of Words—Unmistakably Ranulph

This is an extract of a conversation between Ranulph and myself (Glanville, personal communication 26 April 2013).

Philip: Which family therapy is an embodiment of second order cybernetics? My experience shows that mostly they are barely an embodiment of first order! Classical example, Manuchin's application of family therapy neglects his own worldview, his own posture and actions in therapy. Still sees the family as black box…

Ranulph: No. *Application* is a power term. If you apply something to something else, you are controlling it in a linear way.

Philip: Okay I see it's my application of language that is bothering you. You know I am half engineer right….

Ranulph: I'm trying to get you not to use the word *application/apply*. Second order cybernetics is about mutual effects. Apply and application are concepts that make the circle linear.

Philip: I didn't see it like that as I didn't mean it like that. I can now surely see why you are saying its linear. I am going to spend some time thinking about my language and choice of words.

Ranulph: I think we use many words as unquestioned habits. I've not thought about application till very recently, though it has often felt peculiarly strange/difficult to me. It took me years of quiet contemplation to understand why, and to understand the difficulty this brings up.

If I go back to the start of second order cybernetics, what Mead is asking for is consistency: that we act what we describe, we live the story we tell. I only came to put her demands in this framework very recently, and it has released many things for me.

Philip: This is a challenging task. Thanks for your comments. It helps when there is someone to moderate the thinking!

The connectedness versus fragmentation of different language. I am thinking that *application* also alludes to separation. I am here, it is there. I do something to it, or I initiate a change in it.

The paradox of my day, discussing causality with you and re-thinking my language while reviewing a chapter of a new book on Linear Causality and the correlation machine.

One day I'll check into a mental institution and they will ask "what's wrong with this guy" and the nurse will say "he can't decide which hat to wear."

Ranulph: Misconstruction of the question/comment. I'm sure you can decide which hat to wear. It's just that you may not stay with the decision. The misconstruction is that the mental health workers think you must decide once and for all. But the magic of deciding is to keep deciding. If you decide the same and then the same again, you seem to have decided one way for good. If you change, they think you can't decide. The opposite is the truth. What the difference is, is how you make the second decision.

Glanville had a way with words. He knew that conversation was an important aspect of human experience, for conversations are circular in nature with participation a pre-requisite for conversing between parties. Von Foerster reminds us that meaning is not transmitted in the conversation; rather, meaning is what the listener determines from what they heard (Glasersfeld, 2007). While conversations are circular, they also are detached. As each party dialogues, their interpretation at each moment is personal and can only be re-represented to the other through the available understandings of the other. The parties to the conversation are not machines. They are psychological individuals that have emotions, beliefs, ideologies and so forth. Pask (1973) refers to individuals as p-individuals (psychological). For some uniformity in understanding between parties of a conversation, Glanville advocates the principle of mutual reciprocity. Glanville states:

> The Principle (or Law) of Mutual Reciprocity states that, if through drawing a distinction we are willing to give a certain quality to that we distinguish on one side of the distinction, we must also permit the possibility of the same quality being given to that which we distinguish on the other side of this distinction: If I distinguish myself from you and I consider I am intelligent, I must consider that you (which I distinguish from I) might also be intelligent …This suggests that generosity of

approach is important. We should look to find and affirm qualities both in another and in ourselves. We seek to welcome these qualities, rather than deny them…I have argued that favouring such positive qualities is a major benefit of second order cybernetics (Glanville, 2004). We can develop a richer account of being human than the impoverishing approach so familiar in the materialist utilitarian interpretations, which assert our essential selfishness, suggesting the model for human behaviour is mean and grabbing. In contrast, I revel in our generosity. (Glanville, 2008, pp. 168-169)

Conversations rest on generosity and trust, for listening is an act of taking what others offer us and interpreting it (Glanville, 2003, 2012). Glanville states:

When we listen we can take onboard (our understanding of) the understanding of others. This gives us a way of transcending our own limitations, of moving beyond what we can imagine. Thus, we can expand our horizons, and open ourselves up to the thrill of the new. What we take from others is not, however, what they give, but is how we hear what they offer. The act of listening involves us in an active and creative act (which is why listening is not just picking up information transmitted). It is therefore our interpretation of what is offered, rather than simply what is offered. (Glanville, 2012, p. 168-169)

The word that stands out to me in this quotation is *how*. Glanville talks about how we hear, eluding to the process of our hearing. I interpret this as his strongest characteristic. A characteristic portraying an ethical stance that evokes his own responsibility in the conversation. Glanville knows that he is responsible for what he understands and how he comes to this understanding. While there is scope for error correction in conversation, clarifying ourselves as we move through this dance—as von Foerster termed it—language is not a code, but only a medium for negotiation in a dialogical space. Language is a set of conventions that has its own rules that we learn as we join the club of those who use it (Glanville, 2012), and we notice that different clubs use the language in different ways. It's a continuing process in which we learn about our self, change, and through mutual interaction we observe our self reflected in the other. This next quote I call Glanville's disclaimer :

Second order Cybernetics is a study in which observes and actors take responsibility for their observations and actions. Therefore it goes without saying that, while I greatly enjoyed being on the receiving end of my colleagues' criticisms, all errors intentional, born of ignorance or opinion, or otherwise created are mine. I do not just accept them. I welcome them: all errors remain the responsibility of their owner! (Glanville, 2012, p. 204)

Conversing with Ranulph through the three years I knew him was challenging. Communicating mainly over email was helpful as I could have time to reflect and think about his perturbations to me. Ranulph said to me that he realised that reflection time was just as important as forward thinking work time. For example, working on a project and being in constant forward moving impetus of "what's next?" must be followed by a time to reflect. The reflection time should be given equal weight. This advice has proved invaluable in my own life and I have found more change in my own thinking arising out of the reflection time than the busy action orientated time.

Referring back to the dialogue presented at the start of this section, something changed in me. It was like the often used saying "the penny dropped." Here I was involved in cybernetics, yet did not see that there was inconsistency between my thinking and acting in cybernetics. I now started to think about how I was communicating. I acknowledge Ranulph's personal interest in me during the time I knew him, which I believe was truthful and generous.

Theory and Practice

Glanville draws a distinction between von Foerster's second-order cybernetics and Mead's (1968) cybernetics of cybernetics. Glanville states:

> It might appear there is no difference between (first and) second order cybernetics, and Mead's cybernetics of cybernetics. We have certainly acted as if this were so. But I believe this is not so, and that the difference between the two is crucial to unravelling some of the confusion in how we describe the system and what we can describe. I have claimed Mead is concerned with consistency, particularly consistency between how we understand and how we act. She requests we act in a manner reflecting our understanding...
>   Heinz von Foerster (1974) (the ring master of second order cybernetics) distinguished second from first order cybernetics in the following way: first order cybernetics is the cybernetics of observed systems; second order cybernetics is the cybernetics of observing systems.
>   And I, again, have used my device of preposition-switch to distinguish them (Glanville, 2005) to talk about the difference of having: an observer OF the system; an observer IN the system.[2] (Glanville, 2015, p. 1177)

Glanville (2015) believes that extensive work has been presented IN cybernetics that is presented by observers OF the system essentially taking the position of first order cybernetics. He goes as far as saying second-order cybernetics has become obsessed with the observer. However, Mead requires us to act consistently and is more concerned with the experience of the observer IN the system and the behaviours thereof. To remain consistent, as Mead challenged us, the observer IN the system needs to be accounted for, not only in their observations but also in their way of behaving. Von Foerster recognised this inconsistency when he insisted that "the Laws of Biology must write themselves" (Foerster, 1972).

It is challenging for newcomers to the field of cybernetics to grasp a cybernetic epistemology. Further, second order cybernetics often requires a complete change in one's understanding of reality with the observer included in the system and the ethics this entails. These are hard lessons for even seasoned academics. Performing reviews on many papers and working with students learning cybernetics, I experience people frequently falling into the trap of observer independent realities, yet still thinking they are acting from a cybernetic view—myself included. This is one of Glanville's strongest themes, his ability to articulate cybernetics even drawing distinctions between what most people deem the same thing—second-order cybernetics and

---

2. The observer IN the system is the second-order cybernetic observer.

cybernetics of cybernetics—and basing this on the context of the early thinkers. It is this attention to detail with the well-developed language toolset that sets Ranulph in a league of greatness. Taking this even further, to move from thinking about one's position in a system, to actually behaving in line with a cybernetic ethic within the system, is a formidable step. This I believe is Ranulph's major contribution to cybernetics. He attempted to attain what Mead eluded to in her paper on cybernetics of cybernetics, while deeply conceptualising the role of the observer. Aligning more than one's thinking to cybernetics, but actually living cybernetics.

> Cybernetics is not just a study, but a way of acting. We live in cybernetics. If we wish cybernetics to regain its former influence, we should consider our way of living in cybernetics as an example that may attract others.
> This, for me, is at the heart of the understanding of what cybernetics is. (Glanville, 2015, pp. 1174-1175)

In what will probably be his final publication, he postulated a return of cybernetics to a subject that focusses on acting as well as understanding, pointing to effective ways of acting (Glanville, 2015).

For people learning cybernetics, a very popular question arises: "What is cybernetics?" Glanville knew what cybernetics is. Looking at his life I can say the following: When there is alignment between your theory and practice and that you behave consistently in an ethical approach cognisant of your own responsibility in creating meaning, enlightenment may be achieved, and the answer as to what cybernetics is, will be available to you.

Habitual Behaviour

In the conversation presented earlier, Ranulph stated, "I think we use many words as unquestioned habits" (Glanville, personal communication April 26, 2013). Cybernetics being intertwined with ethics, forces one to question one's own way of understanding, one's own way of behaving, and how this behaviour is experienced by others. Whether these habits relate to addiction, insincerity, ignorance or any other human action, examining our own way of being is an ethical imperative associated with second-order cybernetics. This differs from morality, as von Foerster (1992) believes ethics are the personal tangle between one's own beliefs and values, which is different from morals, which are applied by others to others (von Foerster, 1992). This makes cybernetics a very personal endeavour. Taking time out to simply reflect and conceptualise one's unquestioned doings and sayings—heuristics—can be sobering. In doing so in my own life, I have become more aware of pre-conscious actions that I was not aware of. These actions were defining me, which were not at all a best practice that I wanted to live by. If someone held me to some of my behaviours, I would choose to do them differently if I had another chance, meaning they were not aligned to my theory of living in the first place.

In theory, theory and practice are the same. In practice, they are not.[3] Breaking these ill-conceived mental heuristics, breaking faulty patterns, and aligning theory and practice rely on considering the antithesis to the thesis. Every person has their unique way of making sense of their world, which is like their thesis. In cybernetics, Glanville constantly advocated an investigation into both the thesis and the antithesis. By evaluating the antithesis, options are created, which in itself is a good cybernetic step, keeping in mind von Foerster's ideas about acting to increasing options.

Conclusion

I disagree with Gary Boyd as his critique was one sided. Saying Glanville's work lacked a theory without acknowledging theory and praxis as equally important, tells only part of the story. Just as Aristotle spoke of sophia arising from phronesis but also returning to phronesis, so too must one view all aspects of a person's contribution before discounting or highlighting errors. Ranulph's 2013 ASC conference theme embodied this theme "Acting, Learning, Understanding" (Glanville, Baron & Griffiths, 2014). Glanville (in press) felt that there should be no superiority with acting neither more nor less important than understanding. In the same token, theory is no more superior than practice. Glanville states:

> The argument I am making is not intended to replace understanding (theoretical knowledge, sophia) by acting (practical knowledge, phronesis), but to redress what I have shown to be a damaging imbalance promoted by the dominance of theory, by emphasising a complementarity between these two styles of knowing. (Glanville, 2014, p. 1298)

The key here is that for a critique of Glanville to hold merit, one would need to place it in the context of Glanville's last ASC conference theme of "Living in Cybernetics". Here the how of one's theory and understanding is brought into the picture. We can ask, how does this person live a cybernetic life or at least how are they aspiring to this way of being? Moving from a *what* or a *why* to a *how* changes the view. From this stand point, Boyd's critique shows that the other side of Glanville was left out. Saying Glanville lacked a theory tells more about the observer who made that distinction rather than the thing observed. Observing a person and punctuating an error in their existence is in itself troubling, for it is the observer who sees this error and thus shares in its creation. An alternative critique may be to investigate whether Glanville's (1988) thesis (Objekte) and associated works align with his way of being and acting. This of course with the person IN the system accounting for the observer IN (Glanville) and OF the system, keeping in mind that all these observers are p-individuals.

Glanville, taking Mead's cybernetics of cybernetics personally, aligned himself to acting to understanding and understanding to act, while living in his cybernetics. Glanville aspired to the ethical imperative set out by von Foerster (1992): "If you

---

3. Believed to be a quotation by Einstein.

desire to see, learn how to act." This is just it, Glanville completed the circle by exemplifying a consistent way of being cybernetic [stated from the position of one observer—or in memory of Ranulph—one source of trouble].[4]

References

Foerster, H. von. (1972). Responsibilities of competence, *Journal of Cybernetics, 2*(2), 1-6.

Foerster, H. von et al. (1974). *The cybernetics of cybernetics*. Champaign-Urbana, IL: Biological Computer Laboratory, University of Illinois.

Foerster, H. von (1992). Ethics and second-order cybernetics. *Cybernetics and Human Knowing, 1*(1), 9-19.

Glanville, R. (1988). *Objekte*. Berlin: Merve Verlag.

Glanville, R. (2003). Behaving well. In I. Smit, W. Wendall, & G. Lasker, G (Eds.), Cognitive, emotive and ethical aspects of decision making in humans and in AI. Windsor, Ontario: International Institute for Advanced Studies in Systems Research and Cybernetics.

Glanville, R. (2008). A cybernetic musing: Five friends. *Cybernetics and Human Knowing, 15*(3-4),163-172

Glanville, R. (2012). *The black box: Vol. 1. Cybernetic cycles*. Vienna: echoraum.

Glanville, R. (2014). Acting to understand and understanding to act. *Kybernetes, 43*(9/10), 1293 - 1300.

Glanville, R., Baron, P., & Griffiths, D. (Eds.). (2014). Acting, learning, understanding [Special issue]. *Kybernetes, 43*(9–10).

Glanville, R. (2015). Living in cybernetics. *Kybernetes, 44*(8/9), 1174-1179.

Glasersfeld, E. (2007). The Constructivist View of Communication In A. Müller & K. H. Müller (Eds.), *An unfinished revolution?* (pp. 351-360). Vienna: echoraum.

Mead, M. (1968). Cybernetics of cybernetics, In H. von Foerster, J. White, L. Peterson, & J. Russell (Eds.), *Purposive systems* (pp. 1-11). New York: Spartan Books.

Pask, G., Scott, B. & Kallikourdis, D. (1973). A theory of conversations and individuals (Exemplified by the Learning Process on CASTE). *International Journal of Man-Machine Studies, 5*(4), 443-566.

---

4. Glanville was looking for a new word to replace *observer.*

Bunnell, P. (2015). *Not Pen and Ink*. Un-retouched photograph.

Cybernetics and Human Knowing. Vol. 22 (2015), nos. 2-3, pp. 155-156

My Time with Ranulph Glanville

Lily A. Fischer[1]

In this paper I recall some memories I have of meeting Ranulph Glanville. I examine the question of whether Ranulph was a serious and tough professor. Based on my memories of him I conclude that Ranulph could be very gentle and sweet.

My Time with Ranulph

I have met Ranulph a few times. He was a professor at several universities, hard working and known all over the world. Some might therefore think that Ranulph was serious and tough. I examine this here with recollections I have of spending a little time with him on a few occasions.

I met Ranulph for the first time in Olympia, Washington when I was three years old. I had heard about him before. He was friendly and introduced himself as Ranulph. He explained to me that his name comes from an old Scandinavian word for "to reign", and that he was, therefore, "King Ranulph." Then he roared like a grumpy King.

Three years later, in early 2012, Ranulph visited us in Suzhou, China on his way back from Australia to England. As a gift he brought me a plush kangaroo with a baby plush kangaroo in its pouch. Ranulph named them Bluey and Pinky. I still have them. Later that year, I met Ranulph again at a conference in Asilomar, California. I found most of the conference boring and preferred to play outside. Somebody had to have an eye on me, especially on the beach and near the water. Ranulph volunteered to be my baby-sitter for an afternoon and took me to the beach where he sat down and I played with sand and with kelp that had been washed ashore. We were happy and had a relaxed chat, but I forgot most of the details. As we came back it was already dark. I was no longer wearing my shoes. I had to take them off after a lot of sand had gotten inside of them. I was afraid that people back at the conference would laugh at me for returning with naked feet. But Ranulph said that he would defend me if that would happen. It did not happen but I was glad Ranulph was on my side.

When my parents and the other ASC officers had their last Skype meeting with Ranulph in late 2014 I was already in bed, and my parents made a recording. At the end of the meeting my mom conveyed my best regards to King Ranulph. He was happy about this and roared once again like a grumpy king. Then he said good-bye, and his voice was very emotional.

---

1. Suzhou Industrial Park Foreign Language School, Suzhou, China. Email: lilyfischer@asia.com

Conclusion

I have shown in this paper that, whenever I met him, Ranulph was very nice and sweet. Sometimes, but not too often, he was even a little bit silly.

Acknowledgments

My parents helped me write this paper.

Figure 1: Ranulph and I head for the beach at the Asilomar Conference Grounds.
(Photo: Christiane M. Herr)

In the moment of conception

We perceive

the conversation...

Cybernetics and Human Knowing. Vol 22 (2015), nos. 2-3, pp. 157-166

Virtual Logic—Laws of Form and the Mobius Band

Louis H. Kauffman[1]

Your inside is out and your outside is in.—Beatles, 1968

This [paper] examines the grounding of George Spencer-Brown's notion of a distinction, particularly the ultimate distinctions in intension (the elementary) and extension (the universal). It discusses the consequent notions of inside and outside, and discovers that they are apparent, the consequence of the difference between the self and the external observer. The necessity for the constant redrawing of the distinction is shown to create "things." The form of all things is identical and continuous. This is reflected in the distinction's similarity to the Mobius strip rather than the circle. There is no inside, not outside except through the notion of the external observer. At the extreme, the edges dissolve. The elementary and the universal thus re-enter each other. "Your inside is out and your outside is in."—Glanville & Varela, 1980, p. 638

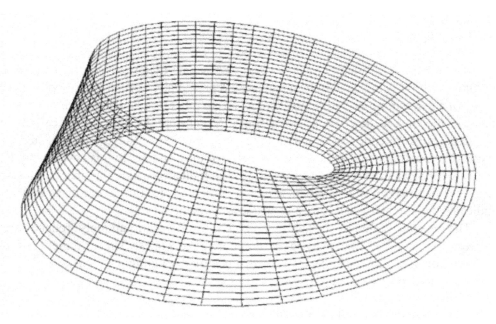

Figure 1 – The Mobius Strip

I. Introduction

This column is devoted to a discussion of the themes and particularities of a paper written by Ranulph Glanville and Francisco Varela, entitled "Your Inside is Out and

---

1. Email: kauffman@uic.edu

Your Outside is In" (Beatles, 1968 quoted in Glanville & Varela, 1980). See the quote above and the image of a Mobius strip in Figure 1.

We begin the discussion by consulting with Cookie and Parabel. Cookie and Parabel are a pair of sentient text strings who claim to be very close to the void, as their existence depends entirely on the minds of the readers of the text strings that they are. Cookie and Parabel have appeared before in the virtual logic column with (Kauffman, 2004) their first appearance.

Cookie and Parabel are quite at home discussing the calculus of indications of G. Spencer-Brown's *Laws of Form* (Spencer-Brown, 1969). I am sure that many readers of this column are also familiar with laws of form (LOF), but here is quick reprise of Spencer-Brown's calculus of indications.

The calculus of indications is based on a single symbol, < >, called the mark. We take the mark in typographical form as shown in the last sentence for convenience in writing. The mark is seen to make a distinction between inside and outside, just as the double angle bracket < > is used to enclose what is on its inside and to disclose what is not on its inside. One has a calculus of distinctions based on the mark and based on two rules:

Calling: < > < > = < >.
Crossing: << >> =   .

In the rule of crossing, we have replaced a pair of nested empty brackets by nothing. Calling is intended to be a symbolic representation of the law of calling: The value of a call made again is the value of a call. In other words, the mark is seen as the name of the marked state. Then two calls of this name are the same as one call of it.

Crossing is intended to be a symbolic representation of the law of crossing: The value of a crossing made again is not the value of the crossing. In this case we say that the value indicated by a mark is NOT the value indicated inside that mark. Thus < > indicates the marked state since its inside is unmarked. And the outer mark in << >> indicates the unmarked state since its inside is marked.

The ensuing calculus is subtle and simple. We can uniquely reduce finite expressions to either the marked or the unmarked states. The arithmetic of the Calculus of Indications can be interpreted for logic by taking <> to denote True (T) and <<>> to denote False (F). Then

<A>B models "A implies B".
AB models "A or B".
<<A><B>> models "A and B".
<A> models "Not A".

It is easy to see how all this works. For example <A> is marked exactly when A is unmarked. <<A><B>> is marked exactly when both A and B are marked. Try it!

In the discussion to follow of Cookie and Parabel some remarkable progress is made in understanding the basis of the logic of *Some* and *All*, that is, the logic of syllogisms. In LOF Spencer-Brown has a remarkable analysis of this logic based on the principle that understanding Some and All is a matter of understanding that place where the universal (All) and the particular (Some) are concurrent. Cookie and Parabel begin their discussion with the Mobius theme in the paper by Ranulph Glanville and Francisco Varela. These authors make the point that adding a Mobius quality to the distinction boundary brings us into that realm of the universal and the elementary (their term for the particular). Cookie and Parabel discuss the Mobius distinction and find an arithmetic and an algebra for it. They then discover that this Mobius Calculus is just what is needed to articulate Spencer-Brown's work on the concurrence of the Universal and the Particular in Logic. I heartily recommend this discussion to the reader and I am personally very grateful to Cookie and Parabel for letting us eavesdrop on their wonderful insights.

II. The Discussion of Cookie and Parabel

Parabel: If we are going to discuss that complex paragraph of Ranulph and Francisco, we had best clarify what is George Spencer-Brown's notion of a distinction.

Cookie: I can do that in a nutshell. Spencer-Brown wrote a book called *Laws of Form* in which he shows how the idea of distinction gives rise to all the forms of knowledge that we have. He says "We take as given the idea of distinction and the idea of indication and that one cannot make an indication without drawing a distinction. We take therefore the form of distinction for the form" (Spencer-Brown, 1969, p. 1)

Parabel: I remember that. That is one of the best text-strings that I have ever encountered. It reminds us that any text string is made from the indications of its letters and its words and its meanings. It reminds us that whatever is meant by form, anything that we call a form is an indication and so makes a distinction. The concept of form is a concept of distinction and he says it best with the sentence, "We take the form of distinction for the form" (Spencer-Brown, 1969, p. 1).

Cookie: I am looking at this quote from Glanville and Varela. They say that notions of inside and outside depend on both the self and the external observer. Are you an external observer of my text string? What about external observers?

Parabel: Well Cookie, let's talk about what Spencer-Brown does. He makes an injunction: "Draw a distinction!" And he illustrates this with the drawing of a mark in the form of a right angle bracket. For my purpose I can make a sign that serves the same purpose by writing < > for the mark. You will notice that my typographical mark distinguishes an inside, < Inside >, from an outside, Outside < > Outside.

Since this text produces a one-dimensional space, the Outside got cut into two pieces by our mark. Spencer-Brown avails himself of the plane and as a result can have a connected mark so that both the inner and outer regions are connected. But yes, the mark < > makes a distinction from the point of view of a reader of the text, between inside and outside.

Cookie: Oh! Now I see. I am standing outside the mark as an external observer, and I see that it (the mark) makes a distinction between inside and outside because it makes a wall that I cannot penetrate to get to its inside.

Parabel: Spoken like a true text-string, Cookie, but I am a bit puzzled about what you can do. You are, after all, only one-dimensional. But you can also act on text strings to find out their contents. Do you sequentially read the letters?

Cookie: Yes. So if you have a mark < >, then I can parse it, and I do this from left to right as I was trained. So I produce the string "Left Pointer, Right Pointer" when I "see" the mark. You see I produce another string from the mark. And I realize that this creates an Inside between Left Pointer and Right Pointer where any other string could be inserted. And the Outside is what is to the left of Left Pointer and what is to the right of Right Pointer.

Parabel: When you read the mark why don't you write < > instead of Left Pointer, Right Pointer?

Cookie: What a good idea. And then I would be the mark, since I am the strings that I write.

Parabel: You have got it! Spencer-Brown writes at the end of *Laws of Form*, "We see now that the first distinction, the mark and the observer are not only interchangeable, but, in the form, identical" (Spencer-Brown, 1969, p. 76). Being a text-string has its epistemological advantages. You are text and the mark is text, and so, in the form, you and the mark are identical. It is so simple.

Cookie: Yes. I imagine there are many exalted beings who think that they are different from us, that they transcend text. I was speaking to a paragraph not long ago, and he declared that he was superior to a sentence and that he had nothing in common with a mere mark or letter or even a word.

Parabel: Ok. But now let's see about this mark < >. We both stand outside the mark AND we are identified with the mark. The mark refers to a distinction external to itself and it also refers to the very distinction that it makes. The mark is, as an observer, outside itself. The mark is, as a distinction, inside itself.

Cookie: What about the fact that the mark seems to make a perfect and clear distinction in the line of text? Here is the mark: < >. We are clear that there is a distinct inside and a distinct outside. What is outside the mark is marked. The inside of the mark is unmarked. There are two sides to a distinction, the marked side and the unmarked side. What is not marked is unmarked. What is not unmarked is marked. I say all this and I feel like a barking dog: mark, mark, mark. The value of a bark made again is the value of the bark!

Parabel: You see how that stance leads directly to the two-valued logic. I do not deny it. But once we examine the role of the observer in all of this, then we see that the intertwining of the observer with the mark, and the ability of the observer to also be the observed produces another level where there is no longer such a sharp distinction. It was a creation of the observer that the mark was precise and two-valued. Even speaking of the mark itself takes us to a third realm. For the *edge* of the mark is sitting on a fence, and is neither inside nor is it outside.

Cookie: Hmm… Now you have me thinking in variations. These fellows Glanville and Varela are suggesting that a distinction is more like a Mobius strip than a circle. They are suggesting we think about not only how the apparent distinction of the two sides disappears on the band, but also that the circular band itself is a better model of a boundary. The boundary is a *Mobius Fence* that is penetrated by an inhabitant of the inside who wishes out, or by an outsider who wishes in. Just so, I am distinct from the world, and yet I enter and leave that distinction at will, becoming entwined with the world and then again retreating from the world.

Parabel: The insider who wants out must move onto the Mobius strip and then slide through the twist. He will come out all right, but upside down! I don't suppose he will mind. He'll just pick himself up and continue on.

Cookie: I could make a Mobius Calculus of strings. It would go like this.
    <Up> = < > Down and <Down> = < > Up.
Down and Up and the states of the observer inside of outside the boundary. If he wants to move through the boundary, he will have to flip from Up to Down or from Down to Up. The Mark < > will be a Mobius operator.

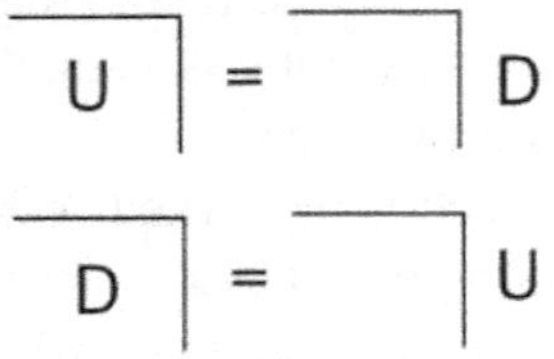

$$U\ | \ =\ | \ D$$

$$D\ | \ =\ | \ U$$

UD = DU = S
UU = U
DD = D

Figure 2 – The Mark as Mobius Operator.

Parabel: Hmm… Lets see. Then we would have
    <<Up>> = << > Down> = <<>>Up = Up.
That makes sense. And
    <Up> < > = < > Down < > = < > < Up>.
Also ok. I like this calculus. What can we do with it?

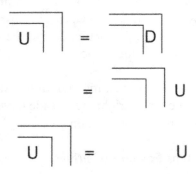

Figure 3 – Examples for the Mark as Mobius Operator.

Cookie: Lets make it more symbolic. Lets use U for Up and D for Down. Then we have
    <U> = < > D. and <D> = < >U.
These entities U and D move through distinction boundaries at the expense of being turned upside down.

Parabel: There is the value S = UD = DU that is the same up or down. Then
    <S> = <UD > = <U>U = < > DU = < > UD = < > S.

This guy S can slip though any net of distinctions without getting flipped. S is kind of like a neutrino. We seem to be talking about tunneling through a barrier. The U and D can tunnel but turn into each other.

Cookie: Lets go back to Mobius calculus using the Mark. We can view U and D and their combination S = UD as a new sort of variable in relation to the arithmetic. It makes sense to say that UU = U and DD = D. Then we have three values: U, D and S = UD. We do not assume that X < > = < > as we would if X were standing for the pure marked or pure unmarked states. So we have a calculus that satisfies the rules
    <<X>> = X for any X and XX = X for any X.
The three Mobius values make these the basic rules for our calculus. We could call these the Looking Glass Rules!

But wait a moment. I am getting worried here. I think we have made a significant addition to the calculus of indications by allowing the Mobius values into the play. The Mobius values are a way to indicate the parity issues of the calculus. If a U appears outside an expression, then it can be transported to any depth in the expression that has the same parity. That is, if you transport it across an even number of boundaries than it will appear as a U. Similarly for D.

Also the Mobius calculus gives us a way to think of the expression as a sort of house with rooms that can be visited by the Mobius variables. Each boundary is a looking-glass boundary that inverts a Mobius value when it travels through the Mobius mirror. It is a different view of an expression.

Parabel: It is quite a different view of an expression. From the point of view of the Mobius variables themselves, they are travelling about inside an expression. From our point of view, we are watching how they change as they move across the Mobius bands in the expression. And yet we are also Mobius variables in a larger encompassing expression.

Cookie: But what about the Glanville and Varela notion of the confluence of the elementary and the universal?

Parabel: I prefer the particular and the universal. Particulars are elementary as they are composed of some few distinctions and can be grasped even by simple text strings. Now this profound business about the concurrence of the universal and the particular occurs indeed when you go 'round the Mobius band and find that you have been flipped into your opposite. It is then that you realize that the particulars of the distinction you thought was so continent are in fact not really different from one another. You come back to your senses as you lose that distinction. But there is more to it. This calculus with its weaker axioms
    XX = X, <<X>> = X,
is at the concurrence of the universal and the particular in logic!

Cookie: Oh! I know what you mean. It is the Mobius calculus that is the right context for working with syllogisms. Reasoning about ALL and SOME is right at the boundary of the universal and the particular. When we negate ALL we bump into SOME and EXISTENCE. Just so if I say "Not all pigs have wings." This shifts from the universal All to the Particular "Some pigs do not have wings."

Parabel: Lets look at it in the form.
    We use <X>Y for "All X are Y."
    We use <<X>Y> for "Not all X are Y",
and we translate this to mean
    "Some X are not Y".
Then we need to see when syllogisms are valid. So we write

All X are Y.
All Y are Z.
All X are Z.

as the exemplar of a valid syllogism. This translates into
    <X>Y and <Y>Z implies <X>Z.
Which transcribes to
    < <X>Y and <Y>Z > <X>Z.
This (using "A and B" = <<A> <B>>) transcribes to
    <   < <<X>Y> <<Y>Z>> > > <X>Z.
Now allow yourself to reduce this in Mobius Calculus. You get:
    <<X>Y> <<Y>Z>> <X>Z.
In other words, a syllogism with form

P
Q

R

is transcribed as
    <P><Q>R.
If this syllogism is valid then so is any version of it obtained in Mobius calculus. Thus for example we can write
    <P><Q>R = <<R>> <Q> <P>
corresponding to the syllogism

Not R
Q
Not P.

For example, we have that the following is a valid syllogism.

Not all X are Z.
All Y are X.
Not all X are Y.

This is the same as

Some X are not Z.
All Y are Z.
Some X are not Y.

There is no way to permute or substitute variables to move an invalid syllogism into the standard valid syllogistic form $<<X>Y> <<Y>Z>> <X>Z$. This standard valid syllogistic form is a locus in logic where the universal meets the particular.

If a purported valid syllogistic form cannot be transformed into the standard form in Mobius Calculus, then the purported form is not valid.

Cookie: This is fantastic. I previously thought that All and Some referred to multiplicities and had something to do with the world of many things, and of course these words do refer to that. But your explanation of validity of reasoning does not make use of these multiplicities. It works in the Mobius calculus, at a special level of the form where the fundamental distinction is also a joining across a twist, but at that level there is no vast multiplicity as we imagine there should be if we should talk of All and Some. We are working in a simplicity.

Parabel: It is a simplicity and it is the right simplicity. If one attempts to use the calculus of indications for validity of syllogisms it will not work correctly. This is because the calculus alone is designed to handle exactly one first distinction and no more. One needs a hint of multiplicity and the solution is go into the primordial multiplicity of one or none. One distinction or no distinction at all is the first multiplicity—Two. And only Two can play this game.

Cookie: What would happen if I slipped through my own Mobius boundary? Will I vanish?

Parabel: Lets try that and go meet the Cheshire Cat.

Cookie: I sense a mad tea party in the offing.

Parabel: We will have to vanish from this place in the Expression. Let us go together.

III. Epilogue and Beginning

This essay is dedicated to the memory of Ranulph Glanville. Ranulph often talked and wrote about the relationship of the Mobius band with the concepts and experiences of

Spencer-Brown's laws of form. The key point about the Mobius it that it only appears to have two sides. There is only one side to a Mobius band, and so "… the inside is the outside is the inside is the outside….".

Each apparent appearance of form, each bringing forth a boundary into the world is fraught with the Mobius. The boundaries are permeable. They are more than permeable. As we travel them we are transformed from the inside to the outside to the inside … and it is only in fixing a time or a place that we can say that here is the inside, there is the outside. Doors and walls and all of our careful constructions of perfect containment are always shifted and passed through in the course of time.

References

Beatles (1968). Everybody's got something to hide except me and my monkey. *The Beatles (White Album)*. UK: Apple Records.

Glanville, R., & Varela, F. (1980). The inside is out. *Proc. III World Congress of Cybernetics* (pp. 322-324). Conference held December 12-15, 1980 in Acapulco, Mexico.

Kauffman, L. H. (2004). Virtual Logic—Fragments of the Void – Selecta. *Cybernetics and Human Knowing, 11*(1), 99-107.

Spencer-Brown, G. (1969). *Laws of form*. London: Allen and Unwin.

Bunnell, P. (2010). *Dark Luminous*. Un-retouched photograph.

Temporal Architecture

Snowflakes stars flowers
All forms of the same dance
unfolding relations

Patterns of beauty
Making time present
in the dynamic moment

Bunnell, P. (2010). *Self Portraits*. Un-retouched photograph.

Bunnell, P. (2014). *History Revealed*. Un-retouched photograph.

Beauty arises
Unfolding in its form
The one that beholds

Ah…there it is again
The arising – Cosmos
Letting its breath out.

"Breathe," he said and,
in the moment of our movement
The Universe sighed

Cybernetics and Human Knowing. Vol 22 (2015), nos. 2-3, pp. 169-182

ASC

American Society for Cybernetics
a society for the art and
science of human understanding

Second-Order Cybernetics, Radical Constructivism, and the Biology of Cognition:
Paradigms Struggling to Bring About Change

Robert J. Martin[1]

This column is a journey that considers both the failure of second-order cybernetics (SOC), radical constructivism (RC), and the biology of cognition (BoC) to achieve wide acceptance, particularly in science, and the opportunities for SOC, RC, and BoC in the social sciences and other disciplines.

I. Introduction:

For most human beings, the concept of objectivity frames one's view of the world without being aware that it has been so framed. There is a tradition going back to Xenophanes, born in the sixth century BCE, who pointed out that we can't know that we see the world as it is, "for if he succeeds to the full in saying what is completely truth, he himself is nevertheless unaware of it" (Glasersfeld, 1984, p. 25). This has never been popular or a widespread idea. Heinz von Foerster consistently pointed out the idea that we invent the world rather than discover it—a concept that is one of the underpinnings of second-order cybernetics and radical constructivism, and a concept that is supported by Maturana's explanation of cognition in his famous paper, "Biology of Cognition" (1970). Many of us had hoped that the work of Glasersfeld, Foerster, and Maturana would substantiate these ideas in science by providing close reasoning and evidence for them, but this did not happen. My journey began with Heinz von Foerster, BCL, and Herbert Brun, in 1967, and lead to a multi-disciplinary thesis supervised by Brun and Foerster and followed by a career in educational psychology, counseling psychology, and as a composer of thorny art music. The three

---

braided strands identified in the title have continually shaped my teaching, practice as a psychotherapist, and practice as a composer, as well as my understanding of the world and of who we are as human beings. My disappointment is that, over the last 50 years, the overlapping and interlocking ideas of radical constructivism, second-order cybernetics, and the biology of cognition have not resulted in a paradigm shift in the sciences concerned with human knowing, especially biology. The existing and developing insights which characterize second-order cybernetics, radical constructivism, and the biology of cognition are not being incorporated into new research in the sciences, especially in the natural sciences, nor are the insights from these three strands of thinking finding their way to wider audiences who could use them. In this column I consider some of the aspects that have contributed to both negative and positive developments that have been taking place. In considering this topic, we will proceed through seven sections, as follows:

Section II provides a brief description of the overlapping thrust of SOC, RC, and BoC as understanding science and all human understanding as resulting from observations of closed, circular systems (us) who do not have access to a world independent of ourselves.

Section III considers how the way in which science has come to be done in disciplines that function as silos inhibits change in the overall paradigm of science as an exercise in "uncovering a 'real' world" (Glasersfeld, 2012, p. 139).

Section IV takes up the positive side of recent developments by considering the question: What happens when a discipline becomes the focus of its own practice?

Section V considers how, when multiple disciplines come together to work on a topic (as in the Macy Conferences), they forge common language across disciplines, and, in the process, each discipline looks at its own practice.

Section VI takes up the ethical implications of the practice of SOC, RC, and BoC—implications seldom talked about, but which are always just beneath the surface of our practice.

Section VII is a set of concluding statements.

II. Characterizing Commonalities among Second-order Cybernetics, the Biology of Cognition, and Radical Constructivism

The thrust of radical constructivism (as developed by Glasersfeld) is that our knowing is a key that allows us to reach a goal; we cannot perceive or describe or represent the world as it is. Knowing is circular: "The mind organizes the world by organizing itself" (Glasersfeld, 2012, p. 139); that is, we organize the world by organizing our own knowing.

The thrust of second-order cybernetics is that the observer is always a part of the observation. This means that our descriptions do not describe an independent world; they describe a world where our descriptions say as much about ourselves as about what we are describing. Additionally, all systems—physical, biological, social, and mechanical—are circular systems. While this is, in one way, merely axiomatic since circularity is part of the definition of what it means to be a system, it is an extraordinarily useful truism in that it affords looking at complex entities and situations in ways that allow us to investigate their circularity.

The thrust of the biology of cognition, as developed by Maturana in the "Biology of Cognition" (1970) and related papers, is that the nervous system is a closed circular system and cannot apprehend the world directly—whether we are talking about perception in a bee, a frog, a human, or a machine.

For the purpose of the following discussion, we can summarize the assumptions as follows: Every nervous system, in whatever species, is a closed circular system with no direct access to an outside world. Knowing is always circular. Our descriptions do not describe an independent world; they describe a world where our descriptions say as much about ourselves as about what we are describing.

In contrast to these assumptions,

1. Science uncovers a "real" world.
2. Science is objective and independent of personal interests.
3. Truth is discovered through observation and/or experimentation. Discovered truth is embraced, though not always immediately, and becomes part of human understanding. For example, Galileo is regarded as an exemplar of the triumph of rational belief based on scientific discovery over irrational belief based on authority.
4. The practice of science takes the stance that observations are made as if from an objective viewpoint. Among other things, this means that objects and qualities (as identified and measured by instruments or trained observers) are assumed to be part of an objective world, as if the observer has no part in what is identified or measured.

To be sure, Kuhn (2012) poked a few holes in this picture by pointing to the resistance of science to paradigm changes. However, the change we speak of here calls into question the idea that science is an escape from our own biases as observers. It's not that human beings are unaware that all human beings are biased; it's that there is a belief that human beings can escape this bias through science.

We may be reaching a point in our history where belief in objectivity interferes with the ability of science to study itself and exacerbates science's tendency to move all problems into an external realm where they cannot be solved because their connection to usefulness and desirability has been lost. For example, we have no way of approaching regional and world-wide problems in an age of cultural diversity except to fall back on science as the one objective and true thing that everyone must accept—which does not work.

One argument for accepting a biologically founded, constructivist view is that by becoming aware of the limits of our knowing, we can expand science to better study those limits—and our own biases. The point is not to diminish science but to enlarge it to better study itself so that humanity can better study itself and its own knowing.

The remaining sections of this paper develop insights about the difficulties in bringing about paradigm change. The next section considers disciplines as silos.

III. Disciplines as Silos That Inhibit Change

First-order cybernetics fit very well with the goals of power over both man and nature and with the idea of science as objective. First-order cybernetics was relatively easy to accept because, while nibbling along the edges of objectivity and the idea of an objective observer, first-order cybernetics does not require users to deal with the assumption of objectivity in science. The demise of cybernetics at von Foerster's BCL (Biological Computer Laboratory) came when public funding came to an end (Müller, 2012; Umpleby, 2003). Applying for private funding brought, in one case, the response that "the people at BCL did not understand the philosophy of science. They held the conventional view that science involved removing the observer from scientific observations, not paying attention to the observer" (Umpleby, 2003, p. 162).

I cannot image a bigger change in the paradigm of science than a move from the view of science as uncovering a "real world" (Glasersfeld, 2012, p. 139) to the view of human beings as closed circular systems who cannot uncover a world independent of its observers. Keeping this mind, we will not focus on the difficulties of making this specific change, but on the more general forces that inhibit science from taking this alternate view seriously.

A change in the paradigm of what science can know seems to have become almost impossible; or, if it did happen, it would be largely irrelevant because almost everything that happens in science takes place at the discipline level. At the discipline level itself, it is difficult to imagine a paradigm change unless an overwhelming number of practitioners of the discipline experienced a need for change. Even then, it is uncertain as to whether funding agencies would respond positively.

In other words, an individual or small group of researchers might change their paradigm about a specific set of ideas or practices within their discipline; but why would they propose a project or write a paper that might eliminate them from the cycle of proposal, funding, research, and publication—a cycle that in its modus operandi is both normative and conservative? Researchers just don't do anything in their proposals that does not follow established guidelines. In as much as research funding comes from adhering to exactly what the request for proposal calls for, and publication comes from adhering to exactly what is expected by research journals and those who peer review their papers, there is a very strong dis-incentive to depart from the norm. Not surprisingly, BoC has been better received in the social sciences (which are often not funded by outside agencies) where BoC has become a metaphor for social processes, a result that may not have been intended by Maturana but which was, perhaps, inevitable.

The nature of a discipline (or sub-discipline) as exclusionary is nowhere more evident than in its language. One cannot easily read the literature of another discipline or sub-discipline because one doesn't have the background or the vocabulary.

Each discipline develops its own language as a consequence of the need to understand and to be understood in an efficient way. It's easy to criticize this state of affairs, but it is entirely understandable and probably even necessary for the development of a discipline. The development of highly-specialized language allows a specialized group to function efficiently and unambiguously. This is not a matter of how esoteric the subject matter is. In the past, every trade had its own specialized vocabulary. For example, woodworking was highly specialized; and there were terms for all the specialized tools, processes, and products, allowing everyone involved in the trade to communicate efficiently and unambiguously about every part of windows, doors, casings, and moldings. These words have fallen into disuse by the thousands, but similar traditions continue at your local auto parts store and numerous other specialized technologies. The point is that this situation is probably unavoidable; and getting rid of jargon is not reasonable or possible and, therefore, not a solution to the problem of disciplines as silos.

Proposals for change in the paradigm of science as a whole are going to be in a more generalized language that is understandable across disciplines. Because of that, researchers are likely to consider issues such as the objectivity of science largely, if not completely, irrelevant to one's own research and to the functioning of one's discipline—even though they might lead to changes in one's research. In addition, like any closed group, there's likely to be suspicion in regard to communications about the functioning of the discipline that do not come from within the discipline. It may even be easy for members of a discipline to reject outright arguments about problems in that discipline from anyone outside the discipline (for example, someone from second-order cybernetics concerned about the nature of the observer).

A problem with the topic bias in the conceptualization and the doing of science is that there is a built-in bias against bias, as if it is bad and can be avoided or eliminated. In science, bias means not being objective and disinterested (as in not having a personal stake in something). Bias in any system that does not examine its own functioning and circularity becomes invisible. Science has become such a system, hopefully only temporarily.

Questioning the assumptions of science within a discipline—including how the discipline operates—is likely to raise enormous amounts of cognitive dissonance within and between individuals. The greater the cognitive dissonance, the more likely it will be resolved in favor of existing ideas and beliefs. The idea that science doesn't uncover a "real" world creates a maximum amount of cognitive dissonance. It should be no surprise that a threat to existing ways of thinking about and doing science should be seen as absurd, irrelevant, or dangerous. For example, I remember Maturana saying that his colleagues urged him to abandon the work (personal communication) that became the set of ideas known as BoC, or the biology of cognition—especially "Biology of Cognition" (1970) and "Reality: The Search for Objectivity or the Quest for a Compelling Argument" (1988). Maturana explained that he himself, when he first began to develop the ideas that constitute the "Biology of Cognition" (1970), wondered if he might be going crazy. In other words, the greater the cognitive dissonance it creates, the more likely an idea is to be rejected and then avoided.

A human activity that does not consider its own circularity cannot easily be aware of how it operates. But why would a discipline consider how science functions, if science—including one's own discipline—is a method of objective discovery? Even if individuals or groups do become aware of the need to examine the circularity of their own practice, they are likely to have no control over the larger system of funding and publication that makes science what it is at this time. In addition, if there is no need to consider basic ideas about observation and objectivity in order to carry out research programs, then it would seem a waste of time to consider such ideas. In fact, there is likely to be a disincentive to spend time pursuing ideas that make one less likely to be funded or published. This doesn't mean that there isn't a larger need for science to reconsider its assumptions; it's more the case that science has, at least temporarily, lost this ability to examine itself (see discussion in K. Müller, 2012, pp. 77-91).

The basic conclusion of this section, then, is that there is a stalemate between the traditional idea of objectivity in science and the ideas of circularity and non-representational knowing that characterize SOC, RC (including Piaget's psychology of cognitive development), and BoC. A scientific discipline tends to be closed to paradigms that come from outside itself because they will be perceived as irrelevant—or from inside itself if they differ from the current paradigm because they are a threat—no matter how much empirical evidence and/or how impeccable the reasoning for adoption may be (an exception to this might be the general acceptance of the concept of autopoiesis).

Everything in this section has pointed toward the difficulty of examining one's own practice—a key part of changing one's thinking (including one's paradigm). Everything in the next section is pointed toward considering examples of disciplines that have become the focus of their own practice.

IV: What Happens When a Discipline Becomes the Focus of Its Own Practice?

SOC, RC, and BoC include consideration of their own practice within the scope of their practice. In other words, they undertake to include the study of their own circularity as part of their research. Science does not do this, suggesting that SOC, RC, and BoC are outliers. Viewed in terms of the practice of science and technology, this is accurate, hence the failure of these three research projects. However, in the twentieth and the twenty-first centuries, there are areas of practice that have come so far in the exploration of their own possibilities that they have turned to a focus on their own practice as the subject of their practice. To name a few: design, painting, literature, professional reflective practice, music, and mathematics. When considered in the company of these areas of human culture and accomplishment, it is science that is the outlier, lagging behind other areas of human cultural accomplishment, at least in this one respect. The following paragraphs consider three of these areas, plus second-order cybernetics, organized into four exhibits: painting, music, mathematics, and second-order cybernetics. The discussion of each of these is very brief and is designed to suggest connections, not to make a case for such connections. I leave that to others.

Exhibit 1: What happens when the discipline of painting becomes the focus of its own practice?

In the visual arts, painting went through a series of transformations in which the subject of painting turned from representing "what is real" to styles of painting that became more and more abstract. Painting itself became the subject of painting. Eventually, the experience of viewing a painting became the subject of painting. Robert Irwin gave up painting altogether to focus on the viewer's experience as the subject of visual art (Weschler, 2008). He created works of art that were neither performances nor objects but that consisted of subtle changes in environments that changed the viewer's experience of those environments. For example, by putting a single strip of tape at a certain point in an empty art space, he was able to change the viewer's experience of the space. Whereas art has always been the experience of an object (such as a painting) or a performance (such as play or a piece of music), with some of Irwin's work, art ceases to be something external and exists only in the experience of the experiencing person. Irwin turned the subject of painting from the thing being painted to the experience of the viewer.

Exhibit 2: What happens in composition when composing music becomes the subject of its own practice?

In music, composers explored the possibilities of melody, harmony, and rhythm. As music became more and more complex and chromatic, more and more possibilities of tonal music were explored. Even Schoenberg's twelve-tone system was an extension of the exploration of tonal melody, harmony, and rhythm. After Schoenberg, composers began to explore serialization of all aspects of music. Sound itself, divorced from any tonal context, became a subject of composition. Music could be an exploration of sound and ways of organizing sound, irrespective of any other contexts. If listeners heard melody or harmony or rhythm, the experiences were considered to be those of the listener and not necessarily what the composer had composed. For example, Herbert Brun, longtime ASC member and one of my mentors, spoke of composing events of sound rather than the sound of events—in other words, composing events which generated sounds rather than composing musical gestures. When John Cage composed 4'33" (four minutes thirty-three seconds), his famous piece where the performer remains silent and does not play, the intent was not that were would be four minutes and thirty-three seconds of silence; rather it was that the audience would notice that there would always be sounds. regardless of the venue, and that listeners could become aware of these sounds and could hear them as music (that is, organize them as music in their heads because that's what human beings do). Thus, in different ways, composers began to explore the compositional process itself rather than learning their craft and then using it to compose socially recognized musical items such as ballads, movie scores, orchestra pieces, and so on.

Exhibit 3: What can happen in mathematics when the inventing of mathematics becomes a subject of mathematics?

I imagine it looks like certain columns written by mathematician Lou Kauffman, which have appeared in this journal (2004 and 2014). In the Kauffman's column, two literary devices, Cookie and Parabel, consider the circularity of certain mathematical concepts as well as the circularity of their own existence as it arises in the interaction between text strings and readers. Cookie and Parabel (or, indeed, any literary characters) exist neither in the text strings nor in the reader; instead, they come about in the circularity of the interaction of the reader with the text. Kauffman is not only aware of this, he composes his dialogue with this in mind; and he makes this explicit for the reader, pointing out that Cookie and Parabel are text strings that exist only in the minds of those who experience them:

> Their consciousness comes to life only when they are read. And they have a tendency to fall back into the void as soon as they arise. They have much to say about the foundation of mathematics and the foundations of cybernetics. (Kauffman, 2014 p. 71)

Of course all literary devices exist only in the experience of the reader; the reader probably knew that but not until Kauffman drew attention to it. Kauffman is pointing explicitly to the characters of Cookie and Parabel as existing only in us as we read the text. There is also another level here. Cookie and Parabel "have much to say about the foundation of mathematics and the foundations of cybernetics" (Kauffman, p. 71). The obvious meaning is that the platonic dialogue between Cookie and Parabel teaches us something of mathematics and cybernetics. Another possible meaning is that the concepts that Cookie and Parabel discuss are themselves things that arise only in the experience of readers as they read and contemplate the text—as does mathematics itself.

Exhibit 4: What happens when, as a scientist, the place of the observer becomes the subject of one's observations?

One answer is second-order cybernetics. Second-order cybernetics arises from the contemplation of the scientific exploration of cognition. Cybernetics originates in a stance of objectivity that discovers that its stance of objectivity is not defensible. This is an important point: second-order cybernetics ends up with a different stance than cybernetics started with, a change from a stance of objectivity to what Maturana (1988) refers to as objectivity in parenthesis—an objectivity that results from following the procedures of science but within the limitations of cognition that results from being a closed system.

The preceding four exhibits attempt to show that second-order cybernetics is one of a number of examples where a discipline developed to the point where its next logical object of study was (is) its own process (not withstanding that Lou Kauffman is simultaneously an intentional practitioner of both second-order cybernetics and mathematics).

In considering all four exhibits, it is interesting to note that those who seek a degree in painting can expect to be exposed to the history of painting. Similarly, someone who seeks a degree in music can expect to be exposed to the history of music. In both cases, the student may not appreciate the historical development of the practice, but she will have been exposed to it. Not so much in science. Few who study science are required to study the history and philosophy of science; few who become scientists are given the opportunity to be aware of science as a series of historical developments. One of those historical developments is the search for an objective understanding of cognition by scientists who were forced to conclude that the neurophysiology of cognition does not allow for objective knowing.

As this section points out, relatively recently, more disciplines have developed an interest in investigation making their own disciplines the subject of their own practice. In second-order cybernetics, when the neurophysiology of the observer can no longer support an objective view of the world, the observer becomes the subject of the research rather than what is observed. There are connections and patterns here that are worthy of further exploration by others.

The next section takes up questions of exploring one's practice across disciplines.

V. Across Disciplines: Looking At One's Own Practice and the Practice of Others so as to Forge Common Language

Entering into conversations across disciplines, as happened at the Macy Conferences, requires one to focus one's own practice in order to forge a common language in which to converse with others. One probably can't forge new language with which to speak to others outside one's own discipline without become explicit about one's own discipline. Further, conversing about the same topic from the point of view of different disciplines, each with its own problems, language, and methods, requires one to consider other ways of thinking about a topic.

A community of multiple disciplines needs a common language to understand, and converse. SOC, RC, and BoC are examples of transdisciplinary language and thinking. The Macy Cybernetics Conferences (1946 to 1953), often invoked as a golden time for cybernetics, is an example of transdisciplinary work meant to benefit a profession, in this case, medicine. The Macy Foundation was (and is) not in the business of promoting basic research per se; it is in the business of holding conferences that will influence medicine in a positive way. As is well known, the Macy Cybernetics Conference participants were all invited; the participants were all well-established leaders in their fields, and none were trained in cybernetics. They were experts in their own areas, not in one another's areas; so they did not share methods, goals, or even a common vocabulary—as is typical of all silos. Yet they did find ways to muddle through and produce something of value. The Macy Conferences worked as a temporary learning community whose members did not share a common paradigm; but they came together as an interdisciplinary group to learn about one another's paradigms—and, in the process, to probably unavoidably explore their own.

One challenge in the cybernetics community—a challenge put forward the planning committee led by Ranulph Glanville in the webpage invitation to the ASC 50th Anniversary Conference: ASC Cybernetics in the Future" (ASC, 2014)—is to make available to those outside cybernetics the experience of working in a transdisciplinary way.

An advantage of such opportunities is that to present and develop ideas that will be understood, the group must create language that can be understood by the other people in the room who are not in one's field. The result can be conversations and publications that are accessible to a number of fields. The forged language can be understood by people who participated in the conversations and, hopefully, (with effort) by others in the fields represented by the participants.

Transdisciplinary groups have structures and resources that allow us to work to develop ideas, processes, and products across disciplines—but these groups typically exist for a limited time and a limited number of meetings. The Macy Conferences were unusual in that they were held yearly for seven years (1946-1953). Seven years is a long time for support from a single source but a very short time for meetings in disciplines. The challenge is to keep the transdisciplinary meetings going—and I would see the ASC as one example of a transdisciplinary community within the larger community of readers of this journal.

If it is accurate that science has not been very open to transdisciplinary work, it is also true that some disciplines, including some science disciplines, are more open to transdisciplinary work. In this regard, I was very fortunate that my home field of educational psychology had already found a small place for Piaget by 1967, and for the work of Piagetian scholar Ernst von Glasersfeld, whose radical constructivism has roots both in philosophy and in Piaget's work. Piaget's work is also very compatible with Foerster's ideas about observing systems (otherwise known as second-order cybernetics). My simultaneous association with these transdisciplinary scholars was purely accidental and fortuitous; even so, learning to understand them was a tough slog and could not have continued to happen without being part of a number of communities where these related but diverse ideas, each presented in its own language, could be learned. The opportunity for this kind of learning to take place was what Ranulph Glanville wanted to create through conversation-based conferences. I didn't always see his point, but through the work involved in writing this paper, I am seeing it much more clearly. We can continue to undertake the messy business of becoming a learning community that reflects on itself through conversation-based conferences. This is a very unusual thing for a member of science and professional communities to do. In this regard, the separate and related communities of BCI, of the Foerster/Brun led heuristics seminars (which led to the creation of the *Cybernetics of Cybernetics* (Foerster, 1974) that was completed after I had left the group), seminars in music and composition taught by Herbert Brun, and both classes and teaching of educational psychology as part of my doctoral work, all served to make me transdisciplinary—and often left me very confused.

I was very fortunate in that Piaget and even radical constructivism found a place within educational psychology and also within the profession of public education (though a smaller place than I had hoped). This has taken many decades to accomplish, but it happened. It is happening elsewhere and it can happen in still more places. And we can be part of it in a deliberate and intentional way—if we can figure out how to do it. This can be done by choosing to figure out how to do it; doing it; revising, through reflective practice (Schön, 1983); and then doing it better or at least differently.

This section has considered a stance one can take with others to create transdisciplinary communication. The next section considers the ethical stance one chooses to take vis-à-vis others in the process of making cybernetics—including SOC, RC, and BoC—available to a wider group of interested others.

VI. Ethical Implications

My own position is very clear: I embrace science as a way to help us understand ourselves and our world. I embrace a larger view of science that investigates its own assumptions and methods. We are not stuck in an inevitable reality over which we have no power; we are responsible for the world in which we live. This does not mean that all ideas and methods are equal. Every set of ideas and methods may be more useful than another set of ideas and methods in achieving particular goals. We may be at a bottleneck in history in regard to our ability to solve problems of justice, conflict, and the environment/climate that we have created for ourselves. Acceptance of a view of human knowing based on the biology of cognition could help move us forward out of the bottleneck. Acceptance, however, cannot flow from an acceptance of the biology of cognition as a scientific truth. That would be a non-defensible assertion. If science does not uncover a "real" world, then no description of reality is privileged as the right and correct one. If no description of reality is privileged as the right and correct one, then the description of knowing as not having access to a "real" world cannot be privileged either, even if support for this view is rooted in the most rigorous science.

In other words, SOC, RC, and BoC are ethical when they apply their findings to themselves. Taken both individually and together, they suggest that there is no set of logical or empirical findings that can require us to think in a particular way. Belief in objectivity, regardless of how unnecessary it may be to do science (Maturana, 1988), is a result of human experience and of historical developments—and this belief was, and is, useful in creating support and respect for science—and scientists. A belief system does not have to be arguably correct for it to be useful; it only has to work.

When we embrace a set of ideas such SOC, RC, and BoC, it is because, for whatever reason, as participants in these projects, we prefer them. We can never prove that they are true. By offering our ideas to others as an alternative to their existing beliefs, we give others a chance to make choices. If we indoctrinate others as if our beliefs are correct and true, then we privilege ourselves. To privilege ourselves means to place ourselves above others, in a position of dominance—which easily becomes a form of violence, whether intellectually, emotionally, or physically. And we become liars. Heinz von Foerster liked to say, "Truth is the invention of a liar" (Foerster & Broecker, 2010, p. 19). Foerster's argument against truth was that truth leads to violence—and besides, it's unnecessary (Bunnell & Vogl, 2000; Foerster & Poerksen, 2002). Herbert Brun had a related saying, "Your belief makes me a liar" (personal communication). That is, your belief in what I say makes me into a liar—I only tell you what I know, not what is true—I can't know what is true. Concluding, our observations can result in extraordinarily useful descriptions of the world—even though we are unable to uncover a "true" or "real" world.

To conclude, to do anything in the way of indoctrinating others to our views is to invite the same potential for violence as every other view that is taken to be privileged. I can only offer SOC, RC, and BoC as interesting, useful, potentially convincing sets of ideas and tools for seeing, understanding, and acting. I cannot even say "we" because ethics implies the choices one makes oneself, not choices one imposes on others.

The final section attempts to summarize in a way that will reflect on this journey and where it has taken me, in the hopes that it will result in us reflecting on where the journey has taken us and to consider where cybernetics might go in the next 50 years.

VII. Endings and Beginnings

I began this column to express the deep sense of disappointment I have felt over the last several decades but have ended realizing that we need to keep working. Our insights can be useful in development of many different disciplines and activities. We might want to design ways to develop and share these insights. Develop and sharing go together; we cannot share what we know without also developing it in the process.

The method of working my way out of my sense of disappointment has been the application of second-order cybernetics to its own failure—and finding that no matter how I might want it to be, every part of the human world follows its own logic. The development of science seems not to follow a rational path but the path of wandering stream, directed in its way by its environment—an environment it changes over the course of time. The need is there. We are trying to give birth to something new. Albert Müller writes:

> The decline of Western cybernetics (caused by a lack of funding) finally lead to the creation of second-order cybernetics and to a constructivist epistemology in the early 1970s. This might be regarded as a major step in the history of science. (A. Müller, 2012, p. 58)

From the perspective of the sciences, SOC, RC, and BoC may continue to be viewed as outliers. Viewed, however, from a cultural perspective, SOC, RC, and BoC are part of a wider set of professional reflective practices that is willing to examine its own practice as part of its practice. Can we invent ways of involving ourselves in the wider and parallel universe of practitioners who are already incorporating reflection on their own practice? Can we find sweet spots where others would want our participation? Can we find ways to cooperate with others to create educational experiences—including university education grounded in design, cybernetics, and systems science, as suggested by Stuart Umpleby (2014)? These are examples of ways to connect with others that could be developed by crowd sourcing within the SOC, RC, and BoC communities.

Piaget found that all learning is recursive, which means that all disciplines are recursive, including science. Our job is to help make the recursion a conscious and intentional part of practice. What I learned from reviewing the Macy Conferences is this: Our task is not to teach others; our task can be to collaborate with others, and, in the process, all learn, and in the process, all collaborate, and in the process—you've already gotten the point. A joke: How do you end up with a von Foerster? The Laws of Form got it partly right; the baby makes the same distinctions over and over by doing the same thing over and over: Pick up object, put it down; pick up object, put it down. But every time the object is picked up it is different—even if it's the same object. Therefore every action of picking up is the repetition of a new action. That is, the action is the same but not exactly the same. Piaget calls this a schema. And the schema changes and enlarges, and changes, even as it seems to be the same schema. Every time the baby crosses, the crossing is both familiar and something is added. It spirals all the way up from the baby's first action in the womb until it becomes a von Foerster (or whomever) in the present moment. This is the mechanism of learning and doing. Spiralarity: circularity with a twist. It's what intelligent life does.

Acknowledgements

Thanks to Dr. Randall Whitaker for serving as consultant on the paper, especially in regard to Humberto Maturana and the biology of cognition; also to biologist Dr. Suzanne Martin for editorial suggestions.

References

American Society for Cybernetics. (2014). *Living in cybernetics: 50th anniversary conference of the American Society for Cybernetics: ASC cybernetics in the future.* Retrieved 2/24/2014 from http://asccybernetics.org/2014/?page_id=173.

Bunnell, P., & Vogl, B. (Producers). (2000). *Truth and trust: Three conversations between Heinz von Foerster and Humberto Maturana* [Videocassette]. Washington, DC: The American Society for Cybernetics. (Also posted to http://youtu.be/Mc6YFUoPWSI).

Foerster, H. von (Ed.). (1974). Cybernetics of cybernetics: The control of control and the communication of communication. Urbana, IL: Biological Computer Laboratory Report No. 73.38, University of Ilinois.

Foerster, H. von, & Broecker, M. S. (2010). *Part of the world: Fractals of ethics--A drama in three acts* (B. Anger-Daiz, Trans.). Heidelberg, Germany: Carl-Auer-Systeme Verlag

Foerster, H. von, & Poerksen, P. (2001). *Understanding systems: Conversations on epistemology and ethics* (K. Leube, Trans.). New York: Kluwer Academic/Plenum Publishers. (and Heidelberg: Carl-Auer-Systeme Verlag)

Glasersfeld, E. Von (1984). An introduction to radical constructivism. In P. Watzlawick (Ed.), *The invented reality: How do we know what we believe we know? Contributions to constructivism* (pp. 17-40). New York: W. W. Norton & Co.

Glasersfeld, E. von. (2012). Ernst von Glasersfeld's ADM 2010 dinner speech, Part I. In R. Glanville & M. Ashby (Eds.), *Trojan Horses: A rattle bag from the "cybernetics: art, design, mathematics—a meta-disciplinary conversation" post-conference workshop* (p. 139). Vienna: edition echoraum.

Kauffman, L. H. (2004). Virtual logic-fragments of the void-selecta. *Cybernetics & Human Knowing, 11*(1), 99-107.

Kauffman, L. (2014). Virtual logic—the yablo paradox. *Cybernetics & Human Knowing, 21*(4), 69-78.

Kuhn, T. S. (2012). *The structure of scientific revolutions* (4th ed.). Chicago: The University of Chicago Press.

Maturana, H. (1970). Biology of Cognition. Urbana IL: University of Illinois Biological Computer Laboratory Research Report No. 9.0. (Also posted at: http://www. enolagaia.com/M70-80BoC.html as reprinted in (1980) Autopoiesis and cognition: The realization of the living (pp. 5-58). Dordrecht, Germany: D. Reidel Publishing.

Maturana, H. R. (1988). Reality: The search for objectivity or the quest for a compelling argument. *The Irish Journal of Psychology, 9* (1), 25-82. (Also posted at http://www.enolagaia.com/M88Reality.html).

Maturana, R. & Poerksen, B. (2004). *From being to doing: The origins of the biology of cognition* (W. K. Koeck & A. R. Koeck, Trans.). Heidelberg, Germany: Carl-Auer Verlag.

Müller, A. (2012). The death. In R. Glanville (Ed.), *Trojan Horses: A rattle bag from the "cybernetics: art, design, mathematics—a meta-disciplinary conversation" post-conference workshop* (p. 187). Vienna: edition echoraum.

Müller, K. H. (2012). The new science of cybernetics (NSC): Goals, modes, and cognitive utilities. In R. Glanville (Ed.), *Trojan Horses: A rattle bag from the "cybernetics: art, design, mathematics—a meta-disciplinary conversation" post-conference workshop* (pp. 77-98). Vienna: edition echoraum.

Schön, D. A. (1983). The reflective practitioner: How professionals think in action. New York: Basic Books.

Umpleby, S. (2003). Heinz von Foerster and the Mansfield amendment. *Cybernetics & Human Knowing, 10*(3-4), 161-163.

Umpleby, S. (2014). The role of cybernetics in security policy. *Cybernetics & Human Knowing, 21*(4), 79-82.

Weschler, L. (2008). *Seeing is forgetting the name of the thing one sees: Over thirty years of conversations with Robert Irwin* (rev. ed.). Berkeley, CA: University of California Press.

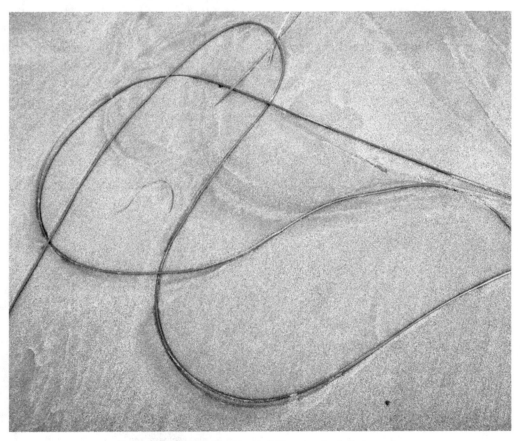

Bunnell, P. (2006). *Not Knotted*. Un-retouched photograph.

Cybernetics and Human Knowing. Vol. 22 (2015), nos. 2-3, pp. 183-187

Two Roads Which Diverged

Phillip Guddemi[1]

The Cybernetics Moment: Or Why We Call Our Age the Information Age, by Ronald R. Kline. Baltimore: Johns Hopkins, 2015.

Ronald R. Kline's new book, *The Cybernetics Moment, or Why We Call Our Age the Information Age*, sees the cybernetics moment as largely in the past, in the period of the Macy Conferences and Norbert Wiener who is posed with equations and a cigar on the cover. But with such a book coming out at this time, it is fair to wonder whether we have finally achieved another cybernetics moment, one in which at least curiosity about the cybernetics movement and what it might have meant—perhaps what it might still mean—has reached a kind of critical mass. Along with other recent books such as Andrew Pickering's *The Cybernetic Brain* and the prolific work of scholars such as Bruce Clarke and Fred Turner, *The Cybernetics Moment* may signify a new interest in a movement which as recently as a decade ago might have seemed to many not only past but nearly forgotten.

Of course this would not have been true for the loyal readership of this journal, and our journal is in fact mentioned several times in the book. We are also in the index, though merely there to be cross-referenced with the American Society for Cybernetics. I will return to how that came to be so.

The main theme of the book is the differentiation, the ultimate splitting, of two potentialities which coexisted during their mutual beginnings. The author calls one of these potentialities cybernetics, the cybernetics which in his view had its moment, only to be eclipsed by the rising star of the other potentiality—information theory, and/or "the information age."

Both of these were born under the shadow of the Second World War and its immediate Cold War aftermath—the 1940s and early 1950s. They were born from the same antecedents and to a great extent from the same mathematics. In the contrast the author draws between the information concepts of Wiener and Shannon—who was, briefly, Weiner's student—he notes that they are using the same equation for information but for a "reverse of sign." In spite of the similarity in their concepts of information, the contexts in which they developed it were sharply different, in ways which prefigured the ways the concepts developed in subsequent decades.

In their famous co-interview many decades later in the *Co-Evolution Quarterly*, Margaret Mead and Gregory Bateson—who both figure prominently in this book—developed the idea of data flowing through the system, in which thinkers are encouraged to have a particular example in mind while developing an idea, lest, in

---

1. pguddemi@well.com

Bateson's words, they "get off into an awful mess of ill-drawn abstractions which act upon other ill-drawn abstractions" (Brand, 1976, p. 38). But Weiner and Shannon had different sets of data flowing through their systems—or to put it differently, they generated their theories using different mental analogies. For Weiner the cybernetic ideas were generated by considering the behavior of, for example, anti-aircraft systems that would predict the courses of aircraft in real time. His cybernetics thus came with a bias towards analog phenomena and adaptation to a changing world. It became the less influential version. Shannon's information theory, on the other hand, was developed with a kind of mental reference to cryptography, and by considering how technology might or might not aid human senders and receivers of ultimately linguistic information. It came with a bias toward digital, discrete phenomena and toward fidelity to the intention of a communicator, and used a lot of symbolic logic. It became the more influential version.

Kline's description of the Macy Conferences, which follows his introduction of Weiner and Shannon, is a page-turner. He shows that the question of models was strongly debated there, with Ashby's homeostat being one important model particularly for those in the Conferences interested in biological and social matters. Neither Shannon nor Wiener were limited in their concerns by the dichotomy mentioned above. For example Shannon developed, as a proto-AI device, a "mouse" which ran mazes. The dichotomy was thus not entirely an artifact of the material to which ideas were being applied. It became a subtler matter of emphasis, with feedback being the guiding idea of cybernetics and information, of course, being the guiding term of information theory.

I find the ensuing book somewhat difficult psychologically for reasons which are probably mostly in its favor. Those of us for whom ideas in themselves are important, and who cleave to them in spite of how they might be received by a larger public, can find ourselves annoyed by histories of ideas which do not really emphasize ideas in themselves so much as what those ideas look like, and how they are received by a public which does not fully understand them. But of course ideas such as those of cybernetics and information theory do exist in a social, political, and institutional context, in which appearances and misunderstandings matter often more than the implications inherent in the ideas themselves.

After its consideration of the Macy conferences, this book does seem to leave off consideration of cybernetic ideas in themselves or how they could have been developed, and emphasizes instead a combination of public reception and political-institutional context. If one is looking for the ideas themselves and how they could have been developed—or still could be, one should probably instead turn to Andrew Pickering's *The Cybernetic Brain*. Pickering's book is subtitled *Sketches of Another Future*, but if one considers the time referent of that future to be that of the Macy Conferences, it is Kline's book that goes further to explain why we have had the future we have actually experienced, not the one that could have been.

In the world as it has actually been, cybernetics became something of a craze in the 1950s. There was some hope that it might prove to become a common language between the natural and social sciences, or to get the latter onto a sounder footing. But mostly then, as since, the popular idea of cybernetics got itself conflated with hopes and fears about machines and their place in the human world. Would they take over our jobs? Would they make our lives better? Within the academy there was skepticism sometimes on a disciplinary basis or, by engineers, questioning about what it was that cybernetics really added to conventional views. Within the social sciences and humanities there was an enlisting of cybernetics to bolster ideas that arguably were already being held and which fit the spirit of the times.

Kline deploys a couple of useful expressions from the sociology of science to describe the obstacles that the new way of thinking faced. One of them is boundary work. This is a sociological concept that encompasses the definition of disciplines, but also defines what is considered interesting within a particular field of incipient research. Kline applies it to how the Marvin Minsky group within Artificial Intelligence treated other approaches (arguably including more cybernetic ones), but also how second-order cyberneticians within the American Society for Cybernetics treated other approaches, at least according to one source. The other useful expression Kline uses is legitimacy exchange. This involves linking one's novel inquiry with fields that are already considered established, and the payoff is often grant money. However, there are serious problems with this strategy. It is often in the interests of legitimacy exchange that some of the less appetizing strategies of academic boundary work are undertaken—including those which involve disassociating oneself with the fringe thinkers who wish to associate with a fashionable concept. For legitimacy exchange and boundary work it does not always matter whether these fringe types are obvious cranks or the pioneers of creative thought. People who want scientific legitimacy within the sciences also tend to disassociate themselves from people who want to use their concepts, or at any rate the auras of their concepts, in such fields as the arts including science fiction—and in that way cybernetics paid the price for its early fashionableness, by losing subsequent academic warfare battles because of the impression that it had been a fad.

There is another legitimacy question that Kline's book exposes—that of CIA influence within cybernetics. Since the 1960s anything that may have been touched by CIA influence would be considered retrospectively contaminated by it in the minds of much of the intellectual public. But according to the mindset of the late 1940s and 1950s, it was perfectly within the purview of the CIA to support particular intellectual currents and trends, if the CIA felt these opposed, or even provided an alternative to, Marxism and/or Soviet activity. To this end the CIA supported not only "hard" scientists but even many New York intellectuals (who were otherwise fairly leftist in orientation) and social scientists. Often it funded favored currents of thought via established channels such as the Ford, Rockefeller, and Carnegie Foundations. It also, and this will come as a bombshell, according to Kline provided "the impetus to create the American Society for Cybernetics" [p. 185].

Numerous individuals within cybernetics evidently had CIA ties—Kline especially mentions Warren McCulloch [p. 186]. Kline does not even consider

Gregory Bateson's open and well-known work with the OSS, the CIA's precursor, during World War II, probably because that work was purely anthropology-inspired (Price, 2008). From the point of view of individual researchers and intellectuals in the 1940s and 1950s, a strongly felt anti-Fascism from the war years often transitioned into an equally strongly felt anti-totalitarianism and thus anti-authoritarian Communism in the postwar years. Of course some cyberneticians, such as John von Neumann, were right-wing cold warriors of an Edward Teller complexion; but even those who were decidedly less invested in such views often cooperated with the patriotic national consensus of those years. Interestingly, cybernetics was readily adopted in the Soviet Union, and during the Kennedy years there was talk of a "cybernetics gap" in which the Soviets could overtake the West in applications. It seems this was as overblown as the contemporary so-called missile gap, but the Soviets did adopt cybernetics as a transdisciplinary discourse in some of the ways Western cyberneticians would have liked. There may have been in some quarters an attempt to leverage this into making cybernetics look suspiciously "red"—which of course would reinforce the need for cyberneticians in the West to burnish their own anticommunist credentials.

But none of this marginalized cybernetics from the legitimacy exchange point of view. According to Kline the reinvention of cybernetics as largely a field dealing with social systems, the development of second-order cybernetics (including Maturanan autopoiesis and Bateson's work), and the drying up of academic patronage systems (which took a more discipline centered turn after around 1970), were all factors in diminishing cybernetics' previous prominence, at least in the circles where it had been prominent.

Meanwhile the idea of the information society gathered momentum. Shannon's ideas formed the core of a theoretical application that divorced information from content and could concentrate on the efficiency of its processing. This fed into the development of improved computer technology which also concentrated on processing. Kline's discussion of information during the postwar era relies rather strongly on tracking media and academic usage of the term "information," giving an impression of an increasing hype-to-content ratio connected with the information concept over time. It is almost as if what is being discussed is not so much a way of thinking but an image connected with a vocable, something like a brand. But this is perhaps more a counter-narrative than a criticism, as the strength of Kline's discussion is how he shows cultural biases playing themselves out in the way societies envision information.

Indeed, what Kline shows as developing over this time and under the aegis of this concept is a strain of utopianism that is oddly conventional, one in which the social changes brought about by information technology itself will, without focused or political effort, lead to broadly favorable and democratic states of affairs. In his last chapter, "Two Cybernetic Frontiers," Kline evokes an influential if perhaps forgotten 1974 publication by Stewart Brand, composed of two articles coexisting in a paperback. The first was an interview with Gregory Bateson in which Bateson set forth his complex late-life epistemology and worldview.   The second was an account of young computer programmers playing the ancestors of the Pong style video games that would conquer the subsequent decade. These were the same programmers, in the same institutions, that would develop the ARPANET which became the Internet, and the computer interface and mouse—and even an ancestor of the Ipad.

The subsequent development of cybernetics and information have led to Brand's second article seeming far more prescient than the first. The complexity of Bateson's ideas has not informed the decades following the 1970s, while information technology has changed the world. But Bateson's ideas, like Wiener's, would have challenged what Bateson called the conscious purposes of influential institutions and individuals. Bateson wanted the linear goals and trends of the postwar era to be rethought and reconsidered, and wanted to use cybernetics to help us think about how to do that. Information, on the other hand, is a genie that does what people want, but does not help them separate wisdom from foolishness. It is a tool that does not change the wielder, or that is its promise. Who wouldn't want more effective ways of doing what they want to do? All the better, of course, if by just pursuing their goals they would, by an invisible informational cyborg hand, make the world better in so doing.

Kline's book illuminates some aspect of the history of cybernetics so as to help us think and gain wisdom. Let us hope that such aims can become a fashion and a fad in a new era.

References

Brand, S. (1974). *II cybernetic frontiers*. New York: Random House. (and Berkeley, CA: Bookworks)

Brand, S., (Ed.). (1976). For God's Sake, Margaret! Conversation with Gregory Bateson and Margaret Mead. *Co-Evolution Quarterly, 10* (June).

Price, D. H. (2008). *Anthropological intelligence: The deployment and neglect of American anthropology in the Second World War.* Durham, NC: Duke University Press.

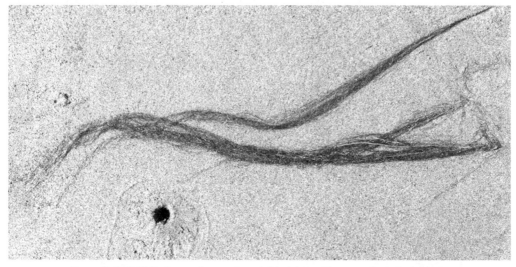

Bunnell, P. (2015). *Remembering Ranulph's Advice*. Un-retouched photograph.

Bunnell, P. (2004). *Golden Canyon*. Un-retouched photograph.

Cybernetics and Human Knowing. Vol. 22 (2015), nos. 2-3, pp. 189-195

The Noninevitable Teleologist

Phillip Guddemi[1]

The Biologist's Mistress: Rethinking Self-Organization in Art, Literature and Nature, by Victoria N. Alexander. Litchfield Park, AZ: Emergent Publications, 2011.

Victoria N. Alexander has an interesting resume. She is a novelist, a biosemiotician, and a complexity theorist who has studied at the Santa Fe Institute. This unique locus of perspective enables her to be sufficiently peripheral to scientific orthodoxy so that she can, almost uniquely, define herself as an adherent to a point of view that has been read out of the theoretical church for generations. This point of view is teleology.

The book under consideration here is a somewhat rambling exploration of what teleology, as Alexander defines it, can contribute to an understanding of natural systems and also of artistic creation. This review is an exploration of ideas and not a point by point summary, and it does not do justice to every room in the sprawling palace of the book. It is a book that reads easily and is full of entertaining stories but the author is not to be underestimated: Its concepts are challenging.

Teleology is usually an accusation rather than an allegiance. It was supposed to have been run over by the locomotive of modern science sometime between its 16th Century formulation and its 19th Century abandonment of theist explanations for biological history. According to the most hard-nosed versions of what is entailed by scientific explanations of cause and effect, these can only utilize only one of Aristotle's four types of cause, namely efficient cause, epitomized by the mechanical action of particle on particle. Final cause, which is what is also called teleology, is the explanation of action in terms of its purposes and goals, which are directed towards the future. This seems obvious in biology: A claw is for slashing at prey, teeth are for biting and chewing, the stomach is for digestion. But the animal whose parts these are, did not design them. Once divine intervention had been removed as a hypothesis for their origination, teleology became in biology an impossibility, while remaining a necessity.

Biology rose to this challenge in two ways. Darwin's theory of natural selection provided the broad brush of a mechanism by which functional adaptation could arise and survive. But for many people it did not provide a full explanation of how biological systems could work in their immediacy. I will in this review supplement Alexander's own analysis at times with what I see as some of its hidden antecedents or forerunners in cybernetics in general, and specifically in the work of Gregory Bateson—but this is not meant to diminish her own work. I see it only as a means to

---

1. pguddemi@well.com

strengthen Alexander's argument for those of us whose backgrounds are more Macy Conferences than Santa Fe Institute.

For the impetus of the Macy Conferences in the 1940s and 1950s was to study, to quote the subtitle of their Proceedings, *Circular Causal and Feedback Mechanisms in Biological and Social Systems*—but these mechanisms were being investigated because otherwise the teleology in these systems' everyday operation was in dire need of explanation. Concepts of circular cause and feedback were felt by early cyberneticians such as Warren McCulloch, Norbert Wiener, and Gregory Bateson to have solved the problem of teleology. And indeed, Wiener suggested that the Macy Conference group call itself the Teleological Society (Kline, 2015, p. 41), echoing the title of a 1943 paper he co-authored.

As the wary son of a heretical geneticist, Gregory Bateson was more circumspect and even made some of the common criticisms of teleology proper in his 1979 book *Mind and Nature*. In Chapter II of that work, ironically titled "Every Schoolboy Knows," he announces that "Causation Does Not Work Backward." The effect cannot precede the cause in time. But his solution, in that space, was the cybernetic one, that of causal systems that become circular.

So of course Victoria Alexander, when she undertook the courageous goal of becoming a teleologist, did so fully aware of teleology's questionable reputation. She writes:

> As J.B.S Haldane is said to have claimed, teleology is like a mistress to a biologist: he may not be able to live without her but he's unwilling to be seen with her in public. The serious and sensible scientists resolutely resist teleology and her meretricious allure. [p. 7]

And indeed there is much that goes by the name of teleological thinking that did not tempt her. For example there is the kind that sees History as having an inherent direction (and inevitable denouement), whether that be toward what was prophesied in religious texts, or toward what was thought to be some inevitable outcome of historical law or tendency, according to secular versions (e.g., Spencer and Marx) of historical determinism. Teleology for her does not involve an external moving force. Rather, and possibly in the Macy Conference spirit, she is concerned with natural self-organizing processes, such as the flying of birds in formation, the similarity of ground plans of unrelated organisms, or the chemical signals that organize slime molds into stalks that perpetuate the species. As an example of a self-organizing system that was originally derived from cybernetic thought, she also mentions Lovelock's hypothesized Gaia, and especially his proof of it using the Daisyworld simulation which demonstrates how an emergent holistic system which constrains its parts comes about without a conscious designer.

For her resolution of the paradox of teleology is mereology. Like Bateson she rejects the causation of the future on the past. But instead of talking about circular cause she posits parts and wholes. For her (italics in the original), *"The whole is the type of result brought about by the interacting parts, and it is the type of result that allows the parts to continue"* [p. 19]. Specifically, *"a flower is made of parts that interact forming the whole, and the whole reciprocally constrains the parts"* [p. 19].

This is an elegant and provocative formulation which would seem to be more general and yet simpler than Maturana's concept of autopoiesis. It is also not foreign to Bateson's mode of explanation, for example in his article in *Steps to an Ecology of Mind* entitled "Cybernetic Explanation," first published in 1967, where he shows that the explanation of the behavior of a cybernetic system—what Alexander calls a self-organized system—is always to be done in terms of *restraints*, which I take to be the same as Alexander's *constraints*. For the outside observer, these restraints/ constraints are clues, or sources of information, which enable an observer to predict an aspect of the system, from knowledge of another aspect, with better than chance results.

It makes one wonder what the difference is, if difference there be, between a *whole* in Alexander's sense and a *context*, where context constrains action taken within it and would enable an outsider to find such action predictable more than chance. Could it be that whenever we see a context that makes a difference for a thing, that is the same as saying we are now looking at a whole comprised by the thing plus its context? Do we generate system boundaries by the simple act of observation, and when we change what it is we are looking at do we thereby envision different system boundaries?

We are headed toward the sign, and semiotics. In Alexander's case she was guided on the highway towards the sign concept by a Santa Fe Institute physicist named Crutchfield, who impelled her to see the constrained behavior of parts as a *sign* of the whole—and/or a *model* of the whole [p. 21].

Crutchfield also made a point that is relevant to the above discussion of prediction and chance. The systems we are dealing with are *nonlinear*, so the behavior of the whole is not understandable simply as a sum of the information one can gather about the parts. That is not how investigation of these systems happens. As Alexander mentions in a footnote citing Crutchfield (1994), "learning to predict the behavior of complex systems involves, not studying the parts and summing up behavior, but interacting with the system such that the observer becomes part of the system and can model the behavior of the whole" [p. 37; The similarity to the anthropological technique of participant-observation is striking.] .

The role of chance in the initial creation of these wholes is an interesting feature of Alexander's discussion. Interesting because even random interaction can form wholes in her sense by constraining the interacting elements which subsequently form its parts. This dynamic could take place among nonliving chemicals, under special circumstances, and of course it is characteristic of life. The whole comes about via processes of "formal and/or functional selection" [p. 32]. And the resulting wholes subsequently adapt, as wholes, to other entities of that type.

Alexander follows the theorist Alicia Juarrero in criticizing Aristotle's original formulation of final cause, which may have put centuries of thinkers off their mark. Aristotle saw goals and purposes as being external to the system, the self or whole. How else could they move the system, if not from an external point from where a lever can be applied? Lakoff and Turner, theorizing basic conceptual metaphors which unconsciously underlie common uses of language, describe the metaphor that seems to have inspired Aristotle as "purposes are destinations" (Lakoff & Turner, 1989, p. 56). But for Alexander, that is exactly what purposes are not. *"Purpose cannot be located at a spatial or temporal difference from the self"* [p. 33; italics in original]

But what is external purpose from a linear analysis looks different from the point of view of what Bateson calls the *self-corrective circuit* (Alexander prefers, possibly for good reason, the term *self-organizing*.) According to Bateson,

> Finally, the famous paper in *Philosophy of Science* by Rosenblueth, Wiener, and Bigelow [entitled "Behavior, Purpose, and Teleology"] proposed that the self-corrective circuit and its many variants provided possibilities for modeling the adaptive actions of organisms. The central problem of Greek philosophy – the problem of purpose, unsolved for 2,500 years—came within range of rigorous analysis. It was possible to model even such marvelous sequences as the cat's jump, timed and directed to land where the mouse would be when the cat lands. (Bateson, 1979, p. 118)

Looking at purpose in this way, native to cybernetics—or as some people think of it, complexity science—Alexander can rethink Aristotle's classical example of a lion hunting a gazelle. As Alicia Juarrero summarizes Aristotle's analysis, a gazelle is an external object which is internally represented in the lion to "actualize the soul's desire" [p. 34] as the final cause or object of desire, motivating the lion.

Alexander supplants this external view of goal with one which internalizes purpose into the circular systems in which the lion is embedded. She is favorable towards the idea of homeostasis and the functionalist logic which it entails, which for early cyberneticians was encompassed in the idea of negative or self-corrective feedback. As I discovered to my chagrin as a young anthropologist, this idea is sometimes considered "conservative"—and Bateson, speaking at a conference of Sixties radicals, marked that there is a conservatism that it always entails:

> Basically these [self-corrective] systems are always *conservative* of something ... always in such systems changes occur to conserve the truth of some descriptive statement, some component of the *status quo* ... natural selection ... may act at higher levels to keep constant that complex variable which we call "survival." (Bateson, 1972/2000, p. 435)

As rethought by Alexander, the teleology of the lion hunting the gazelle precisely tracks this homeostatic epistemology, conservative of the ultimate survival of at least the lion.

> The gazelle is not the real goal. The gazelle is means of survival, which is the ultimate end. So instead of imagining the teleological process going in a linear direction—the lion (agent) chasing the gazelle (goal)—imagine instead the cycle that maintains the lion's life and the gazelle caught up in that cycle. It is as if the lion's selfhood extends itself into its environment and identifies a part of itself (potential food) and takes it in. [p. 34]

And here is Alexander's particular version of biosemiosis. If the behavior of the parts is a *sign* of the whole, and the whole is a conservative or *homeostatic* process, then

> The purposeful acts of chasing, seeking, fleeing and *etc.* are self-organized responses to signs of self...The characteristics of the "prey" becomes a "sign" of the ultimate end of survival when they interact with the predator's evolved repertoire of self-organized responses to the world...The limbs that instinctually begin pursuing the gazelle are not so different from a cell that begins to respond to a nutrient. [p. 34]

But purposeful behavior is not limited to the maintenance of cycles because it is even more purposeful when it is flexible. "All purposeful and/or teleological behaviors should be defined in terms of maintaining cycles that are working well and evolving different cycles when the situation requires it" [p. 35]. When numerous responses are available to a system, we have departed from any strict determinism and begun to work toward a theory of choice.

Even though she has legitimated teleology in terms of survival as the purpose *par excellence*, however, should we see Alexander as one of those who never stop gazing at survival's mesmerizing dance? She has done much of her writing and thinking in response and reaction to postmodernism, and its obscurity. If her foil had been neo-Darwinism, or what is called evolutionary psychology, she might have found herself tired instead of endless explanation in terms of the perpetuation of the individual, the species or the gene. But by the evidence of the multiple topics addressed in this book, she is not what might be called a survival reductionist. Much of the book in fact deals with the arts and literature, and she does not see these as gazelles which only contribute to the ongoing metabolism of their authors.

"No one ever said," she remarks probably inaccurately, "purposes had to be eternal or universal. Indeed, as purposeful activity is synonymous with the activity of aliveness, purposes, like lives, naturally come and go" [p. 201]. The purposes of art, indeed the purposes of daily life, cannot be reduced to a single ultimate final cause.

The key to bridging these is an expanded concept of self. In a discussion of Peirce and the concept of the object of a sign, she returns to survivalism: "only as sustenance or means of survival does the rabbit have meaning for a fox. The object of any sign is always in some sense the continuation of the interpreter; the object of any sign means self, means self-continuation" [p. 88]. But she goes on to note how people will object that human purposeful acts are not necessarily about survival in this narrow sense. People will, for example, choose mates who are not optimum, falling in love with some strange quality in their partner. But their choice arises from the selves they are, at any given moment, comprised of their habits and dispositions as well as their bodies. Again Bateson, in 1960, in the context of a discussion of the roots of emotional pain:

> What is a person? What do I mean when I say "I"? Perhaps what each of us means by the "self" is in fact an aggregate of habits of perception and adaptive action *plus*, from moment to moment, our "immanent states of action." (Bateson 1972/2000, p. 242).

He notes in further discussion that these immanent states and habits are called into being in particular relationships at particular moments. But for Alexander, their continuation in the form of the self they constitute is the Peircean object of the signs they interpret. "Signs are meaningful insofar as they continue a way of thinking or being" [p. 89]. And such ways of being are our selves; we are not mere survival machines.

Indeed, survival is not nearly the main topic of the book, since so much of it revolves around artistic creation which was Alexander's first interest. The main conception of teleology she uses to look at art, such as literature and painting, is self-organization. Indeed she sees the author's role as less hands-on, less dictatorially directive, than many who would not use the concept of teleology in the artistic creation context. Work that is too directional, that points too well to its goal (or moral), is inferior art. It is too predictable and therefore does not fit the Santa Fe criteria for a complex system. Art can originate from observations or components that are at first random, but it is the artist's craft to form these into a whole, or self, which constrains the parts that form it. (Yet this craft is in some sense more like allowing self-organization to take place, and working with the constraints that emerge from it, than it is like placing one's ideas in rows and forcing them to march in formation. Spontaneity, as Alexander points out, should refer to that which is intrinsically caused within a system [p. 44] as opposed to order which is imposed from without. In this sense all good art comes from a process that exploits spontaneity, which is as she says a synonym of purpose not an antonym [p. 58].)

This view of art is surprisingly consonant, as she points out, with Aristotle's poetics. He did not like, for example, the *deus ex machina* resolution of conflicts in a play by the external intervention of the gods. He liked a kind of sense-making which became apparent as the events of the play unfolded, a pattern seen in a Greek tragedy such as *Oedipus Rex*. It is not far-fetched to see this as an example of the creation of a whole that constrains its parts, thus creating (or allowing to emerge) a meaningful unity.

At various stages where Alexander is trying to show how chance or randomness—a vexed topic that she does much to illuminate—can lead to self-organized order, she uses a sort of symbology of L- and T- shapes, which refer mostly to their own structure and possibilities of conjunction. (That which is made possible by structure in this way might be Aristotle's *formal cause*. In the context of living origins it relates to what Terrence Deacon and his co-workers [2012] call *morphodynamics*.) These shapes made me think of the notation of Spencer-Brown's *Laws of Form,* but I am doubtful that they are related to it, though the possibility intrigues me. As these shapes conjoin they sometimes produce "clumps" which have self-continuing properties. Thus the stochastic becomes the systemic via form, process, and time (and is this related to the Peircean concept of taking habits?).

I am going to make short work of long sections of the book on the history of teleology. They are interesting and told well, yet many of the perspectives examined at length epitomize the sorts of teleology that Alexander does not ultimately endorse. These include ways in which analogy is used or misused by people who see in it something like a medieval divine order of providence (such as that which enabled the

Christian writer Boethius to find the consolation of philosophy). Teleology for them involved the scrutinization of the divine will. Freudian treatments of coincidence sometimes put the unconscious in the place vacated by God, or so she argues. History takes a similar divine role for Spencer and Marx (in the Hegelian tradition). And in some ways even the gene, reductionistically understood, plays the part of a *telos*—a fixed goal-purpose—that predetermines futures in ways which actually contradict dynamic systems, cybernetic, or biosemiotic ways of seeing. (On the other hand Kant seems in some ways to have anticipated such forms of thought.)

Nineteenth century determinist science also flirted with teleology but its determinism was called into question by twentieth century developments such as quantum physics. The indeterminism discovered in the last century is to her not teleology's death knell, but is rather a condition for its possibility—when it is understood that it is compatible with radical novelty. Every system will encounter that which is noise to it, which is not part of meaning or significance which it has been able to make. Things out of their contexts do not have "inherent meaning or function" [p. 177]. But they may find new meanings and functions related to other contexts (and other systems). The unintended side effects of actions can provide promise or opportunity as well as danger. Side effects are "like ready-mades just waiting for the right artist" [p. 179]. It seems, in the light of such arguments as these, that teleology is what you make of it. It emerges backwards out of self-created order, which was not strictly necessary going forward. Or was it?

By embracing teleology Victoria Alexander has not eliminated contingency, revived obsolete concepts of transcendent purpose or even reduced purpose to bare survival. She is as fit to encounter a modernist world as any postmodern devotee of contingency and chance, since they have a place in her system too. The end, in the teleological sense, is not found in the final words of a story, or the heat death of a universe. Prior to such final moments, there is plenty of spontaneous self-organization to be *about*—and the *aboutness* which inheres in the resulting self-organizing systems is where Alexander finds teleology.

References

Bateson, G. (2000). *Steps to an ecology of mind*. Chicago: University of Chicago Press. (Originally published in 1972)

Bateson, G.(1979). *Mind and nature: A necessary unity*. Toronto: Bantam Books.

Crutchfield, J. P. (1994) Is anything ever new? Considering emergence. In *Complexity: Metaphors, models and reality. Integrated themes, Santa Fe Institute Studies in the sciences of complexity* (pp. 479-497). Reading, MA: Addison-Wesley.

Deacon, T. (2012). *Incomplete nature: How mind emerged from matter*. New York: W.W. Norton.

Juarrero, A. (2002). *Dynamics in action: Intentionality as a complex system*. Cambridge, MA: The MIT Press.

Kline, R. R. (2015). *The cybernetics moment: Or why we call our age the information age*. Baltimore, MD: Johns Hopkins University Press.

Lakoff, G., & Turner, B. (1989). *More than cool reason: A field guide to poetic metaphor*. Chicago: University of Chicago Press.

Rosenblueth, A., Wiener, N,. & Bigelow, J. (1943). Behavior, purpose, and teleology. *Philosophy of Science, 10,* 18-24.

The tree falls in the wood

And we both hear

the heartwood crack

breaking

the sweet stillness

When hearts break

Why can we not hear

Bunnell, P. (2004). *Spirals Contained*.
Un-retouched photograph.

www.ingramcontent.com/pod-product-compliance
Lightning Source LLC
Chambersburg PA
CBHW080528060326
40690CB00022B/5066